Utilization of Waste-Derived Fuels in the Carbonate Looping Process: Experimental Demonstration and Techno-Economic Assessment

Utilization of Waste-Derived Fuels in the Carbonate Looping Process: Experimental Demonstration and Techno-Economic Assessment

Vom Fachbereich Maschinenbau

an der Technischen Universität Darmstadt

zur

Erlangung des Grades eines Doktor-Ingenieurs (Dr.-Ing.)

genehmigte

Dissertation

vorgelegt von

Martin Haaf

aus Wertheim am Main

Erstgutachter: Prof. Dr.-Ing. Bernd Epple
Zweitgutachter: Prof. Dr.-Ing. Bastian J. M. Etzold

Tag der Einreichung: 14.04.2020
Tag der mündlichen Prüfung: 03.06.2020

Darmstadt 2020

D 17

Bibliografische Information der Deutschen Nationalbibliothek

Die Deutsche Nationalbibliothek verzeichnet diese Publikation in der Deutschen Nationalbibliografie; detaillierte bibliographische Daten sind im Internet über http://dnb.d-nb.de abrufbar.

1. Aufl. - Göttingen: Cuvillier, 2020
   Zugl.: (TU) Darmstadt, Univ., Diss., 2020

   ISBN 978-3-7369-7229-2
   eISBN 978-3-7369-6229-3

Utilization of Waste-Derived Fuels in the Carbonate Looping Process: Experimental Demonstration and Techno-Economic Assessment

Genehmigte Dissertation von Martin Haaf aus Wertheim am Main

Erstgutachter:	Prof. Dr.-Ing. Bernd Epple
Zweitgutachter:	Prof. Dr.-Ing. Bastian J. M. Etzold

Tag der Einreichung:	14.04.2020
Tag der mündlichen Prüfung:	03.06.2020

Technische Universität Darmstadt – Fachbereich Maschinenbau
Darmstadt - D 17

Preface

The Ph.D. thesis "Utilization of Waste-Derived Fuels in the Carbonate Looping Process: Experimental Demonstration and Techno-Economic Assessment" was developed during my work as research scientist at the Institute for Energy Systems and Technology (EST) at the Technische Universität Darmstadt between October 2014 and December 2019.

I would like to thank Prof. Dr.-Ing. Bernd Epple, the Head of the Institute and supervisor of my Ph.D. thesis for the possibility to conduct my own research under his supervision.

Furthermore, I would like to acknowledge the funding from the German Ministry of Economic Affairs and Energy based on a resolution of the German Parliament (MONIKA: "Methanol aus Strom und CO_2 einer Abfallverbrennungsanlage - Untersuchungen zur CO_2-Abscheidung mit Kalkstein" FKZ: 03ET7089) and the SUEZ Group. The majority of the results were achieved in the course of the MONIKA research project.

My special thanks goes to Prof. Dr.-Ing. Bastian J. M. Etzold Head of the Institute for Technical and Macromolecular Chemistry at Technische Universtität Darmstadt, who took over the co-reporting of this doctoral thesis.

Furthermore, I would like to express my deep gratitude to Dr.-Ing. Jochen Ströhle, the Academic Superior Council of the EST for his guidance, support and fruitful discussion. He was always willing to answer any questions that arose in the course of my doctorate.

My sincere gratitude goes to all of my colleagues at the EST that greatly contributed to the success of the experimental investigation at the circulating fluidized bed test facility. Without your work, efforts and ideas, it would not have been possible to achieve these results. Similarly, I would further like to acknowledge the support of the workshop team around Christof and Joachim - even the smartest visions require flanges, steelwork, signals and cables. Thank you Susanne for turning on the light throughout administrative nights. I would also like to thank the students Dennis and Ronak who contributed to this thesis through their student jobs. In addition to all work-related collaborations it was a pleasure to spend everyday-life together with you at the Lichtwiese. Thank you Ralf, Markus, Martin, Maximilian, Peter, Josef, Vitali, Jens, Michael, Jan-Peter, Lorenz, Nicolas, David, Philipp, Christian, Jan, Coskun, Falah, Adil, Adel, Andreas, Dennis, Alexander S., Jochen, Alexander D., Thomas, Nhut, Eric, Carina, Marcel, Ammar, Wisam, Pascal, Ayman, Joachim, Christof, Florian and Waldemar.

Finally, a special thank you to Mr. Rahul Anantharaman from SINTEF ENERGY Trondheim for giving me the opportunity to take part in the NCCS mobility program that allowed me to conduct my research for three months in the beautiful city of Trondheim. Even given the brevity of this stay, I have benefited greatly both, professionally and personally.

Most of all I would like to thank my father Ludwig and his wife Monika for their support throughout the past years. It was so important to have you as a safe harbor. In my thoughts I am with you, my dear mother Christina; unfortunately you can no longer be with us at this time.

Abstract

The economic and environmental threat of climate change due to the anthropogenic rise of the CO_2 concentration in the earth's atmosphere is widely accepted and addressed. Several pathways are being considered in order to diminish the emissions of CO_2 at sufficient mass and time scale. In addition to efficiency improvements, the reduction of energy consumption and the intensive utilization of renewable energy sources such as wind, solar and biomass, the application of carbon capture and storage processes in energy-intense industries seems to be unavoidable.

Through these processes, the CO_2 contained in an exhaust gas stream is separated, purified and subsequently stored in an appropriate long-term storage side. Balancing the CO_2 emissions along the whole process chain, CO_2 reductions of more than 90 % are feasible. The process of CO_2 capture from the respective exhaust gas streams represents the most energy-intensive part of such a process chain which leads to growing research activities in the field of CO_2 capture processes. The carbonate or calcium looping (CaL) process is a second generation CO_2 capture process based on the reversible reaction of calcium oxide and CO_2. The limestone-based sorbent is exposed to cyclic carbonation-calcination reaction regimes, which is realized by means of two interconnected fluidized bed reactors.

In addition to traditional energy-intensive industries such as fossil-fired thermal power plants or the production processes for steel and clinker, waste-to-energy (WtE) plants represent a stationary CO_2 emitter that is reasonably large for the integration of carbon capture and storage processes. WtE plants are predicted to be widely used in future waste treatment strategies, which increases the need to cope with the related CO_2 emissions even further. Due to the organic waste fractions, part of the emitted CO_2 is regarded as carbon-neutral. Once this part is captured and stored, negative CO_2 emissions are feasible, thus CO_2 is removed from the atmosphere.

This thesis evaluates the utilization of waste-derived fuels in the CaL process. In the first step, the feasibility of continuous CO_2 capture by means of a waste-derived fuel fired CaL process was successfully demonstrated by experimental investigations at a $1\,MW_{th}$ pilot plant. In these investigations, the boundary conditions were adapted to an application in WtE plants. Over the course of the test series, it was shown for the first time worldwide that CO_2 capture rates of more than 90 % are feasible, while the CaL process was fueled by commercially available solid recovered fuel.

Based on the experimental data, a CaL process was validated and subsequently applied for the determination of heat and mass balances for the retrofit of a $60\,MW_{th}$ WtE plant. The techno-economic assessment bases on the key performance indicators such as levelized cost of electricity (LCOE) and the cost per avoided CO_2 (CAC). It was found that the LCOE increases from $80\,EUR/MWh_e$ up to $176\,EUR/MWh_e$ in case of the CaL retrofit. This further leads to CAC of $112\,EUR/t_{CO2,av}$. Even though this cost range seems high, it needs to be noted that the application of the CaL process in the framework of WtE plants represents a cost-efficient solution for the achievement of negative CO_2 emissions in comparison to competitive negative emission technologies.

Kurzfassung

Die wirtschaftlichen und ökologischen Gefahren des Klimawandels aufgrund des anthropogenen Anstiegs der CO_2-Konzentration in der Erdatmosphäre sind weithin anerkannt. Es werden verschiedene Ansätze zur Reduktion dieser CO_2-Emissionen diskutiert. Neben Effizienzsteigerungen, der Reduzierung des Energieverbrauchs und der Nutzung erneuerbarer Energiequellen wie beispielsweise Wind, Sonne und Biomasse scheint die Anwendung von Verfahren zur Abscheidung und Speicherung von CO_2 in energieintensiven Industrien unumgänglich zu sein.

Hierbei wird das in einem Abgasstrom enthaltene CO_2 abgetrennt, aufbereitet und anschließend in Langzeitlagerstätten gespeichert. Bei einer Bilanzierung der CO_2-Emissionen entlang der gesamten Prozesskette sind CO_2-Reduktionen von mehr als 90 % möglich. Der Prozessschritt der CO_2-Abscheidung stellt hierbei den energieintensivsten Teil dar, was zu wachsenden Forschungsaktivitäten im Bereich der CO_2-Abscheideverfahren führt. Das Carbonate- oder Calcium-Looping (CaL) Verfahren ist ein CO_2-Abscheideverfahren der 2. Generation, welches auf der reversiblen Reaktion zwischen Kalziumoxid und CO_2 beruht. Das auf Kalkstein basierende Sorbents ist hierbei zyklischen Karbonisierungs-Kalzinierungsreaktionsregimen ausgesetzt, die innerhalb zweier, miteinander gekoppelter Wirbelschichtreaktoren realisiert werden. Der Wärmebedarf des CaL-Prozesses wird durch eine Oxyfuel-Verbrennung von zusätzlichem Brennstoff gedeckt.

Neben den traditionellen energieintensiven Industrien wie fossil-befeuerte, thermische Kraftwerke oder den Produktionsprozessen für Stahl und Klinker stellen Müllverbrennungsanlagen (MVA) einen stationären CO_2-Emittenten dar, der für die Integration von CO_2-Abscheidungs- und Speicherprozessen angemessen groß ist. Es ist zu erwarten, dass Müllverbrennungsanlagen in Zukunft im großen Umfang in Abfallentsorgungsstrategien berücksichtigt werden, was die Notwendigkeit, die damit verbundenen CO_2-Emissionen zu bewältigen, noch weiter erhöht. Aufgrund der organischen Abfallfraktionen wird ein Teil des emittierten CO_2 als klimaneutral angesehen. Sobald dieser Teil abgetrennt und gespeichert wird, sind negative CO_2-Emissionen möglich, wodurch bereits emittiertes CO_2 aus der Atmosphäre entfernt wird.

Diese Dissertation bewertet die Nutzung von Ersatzbrennstoffen (EBS) im CaL-Prozess. Im ersten Schritt konnte die CO_2-Abscheidung aus einem MVA-ähnlichen Abgas durch einen EBS-gefeuerten CaL-Prozess anhand von experimentellen Untersuchungen im 1 MW_{th}-Maßstab erfolgreich nachgewiesen werden. Im Rahmen der Versuchsreihen wurde weltweit erstmals gezeigt, dass CO_2-Abscheidungsraten von mehr als 90 % realisierbar sind, während der CaL-Prozess mit kommerziell erhältlichen EBS befeuert wird.

Basierend auf den experimentellen Daten wurde ein CaL-Prozessmodell validiert und anschließend zur Bestimmung von Massen- und Energiebilanzen für die Nachrüstung einer MVA eingesetzt. Die technisch-wirtschaftliche Bewertung basiert auf wichtigen Leistungsindikatoren wie den Stromgestehungskosten und den CO_2-Vermeidungskosten. Die Stromgestehungskosten einer MVA steigen im Falle einer CaL-Prozess Nachrüstung von 80 EUR/MWh_e auf ca. 176 EUR/MWh_e. Dies führt weiter zu CO_2-Vermeidungskosten von ca. 112 $EUR/t_{CO2,vermieden}$. Auch wenn diese Kostenspanne relativ hoch erscheint, ist zu beachten, dass die Anwendung des CaL-Prozesses im Rahmen von MVAs einen kosteneffizienten Weg zur Erzielung negativer CO_2-Emissionen darstellt.

Table of Contents

List of Figures

List of Tables

Nomenclature

Latin symbols

a	Decay constant of the solid volume fraction in the lean region	$[1/m]$
A	Cross–section area	$[m^2]$
a_1, a_2	Limestone specific deactivation parameter	$[-]$
A_i	Abrasion coefficient	$[wt.\%/h]$
A_{incr}	Incremental abrasion coefficient	$[wt.\%/h]$
A_{tot}	Total cumulative sorbent abrasion coefficient	$[wt.\%]$
b	Limestone specific deactivation parameter	$[-]$
c	Concentration	$[mol/m^3]$
CAC	Cost of CO_2 avoided	$[EUR/t_{CO2,av}]$
d	Diameter	$[m]$
d_p	Particle diameter	$[m]$
d_p^*	Dimensionless particle diameter	$[m]$
e	Specific gaseous emissions	$[mg/MJ_{th}], [g/MJ_{th}]$
E	CO_2 absorption efficiency, CO_2 capture efficiency	$[\%]$
F	Molar flow rate	$[mol/s]$
f_1, f_2	Limestone specific deactivation parameter	$[-]$
f_{active}	Active fraction of Ca-particles in a bed	$[-]$
f_{calc}	Degree of calcination	$[-]$
f_{carb}	Degree of carbonation	$[-]$
f_m, f_w	Limestone specific deactivation parameter	$[-]$
f_l	Average volume concentration of solids in the lean region of a riser	$[-]$
FC	Fuel costs	$[EUR]$
FCF	Fixed charge factor	$[-]$
FOC	Fixed operating costs	$[-]$
G_s^*	Saturated solid mass flow rate in a riser	$[kg/m^2s]$
ΔH^0	Heat of reaction	$[kJ/mol]$
h	Height	$[m]$
HR_{CaL}	CaL process heat ratio	$[-]$
k	Sorbent deactivation constant	$[-]$
k_0	Carbonation reaction rate constant	$[-]$
l	Length	$[-]$
$LCOE$	Levelized cost of electricity	$[EUR/MWh_e]$
m	Mass	$[kg]$
$\dot{m}$	Mass flow rate	$[kg/s]$
M	Molar mass	$[kg/kmol]$
N	Number of complete carbonation-calcination cycles	$[-]$

n	Molar quantity	[mol]
$O_{2,spec}$	Specific oxygen demand	$[kg_{O2}/kg_{CO2,capt}]$
p	Pressure	$[N/m^2]$, [bar]
P	Accumulated distributional sum	$[-]$
P_{el}	Electrical Power	[MW]
PR_{CaL}	CaL process gross power ratio	$[-]$
P_{th}	Thermal Power	[MW]
$\dot{Q}$	Heat flux	[MW]
Q_{spec}	Specific heat demand	$[MJ_{th}/kg_{CO2,capt}]$
r	Interest rate	$[-]$
r_N	Fraction of particles having N cycles between carbonator and calciner	$[-]$
$r_{N,age}$	Fraction of particles having N_{age} full carbonation-calcination cycles	$[-]$
R	Retention rate	[%]
R_{CO2}	CO_2 Recovery rate	[%]
S_0	Initial specific surface area of Ca-particles	$[m^2/m^3]$
S_N	Specific surface area of Ca-particles having N complete carbonation-calcination cycles	$[m^2/m^3]$
SAR	Secondary air ratio	$[-]$
$SPECCA$	Specific primary energy consumption per CO_2 avoided	$[MJ_{th}/kg_{CO2,av}]$
t	Time	[s]
T	Temperature	$[°C]$
ΔT	Temperature difference	[K]
u_0	Superficial gas velocity	[m/s]
u_0^*	Dimensionless superficial gas velocity	$[-]$
u_{mf}	Superficial gas velocity at minimum fluidization conditions	[m/s]
u_t	Terminal velocity of a particle	[m/s]
V	Volume	$[m^3]$
V_m	Molar volume	$[cm^3/mol]$
$\dot{V}$	Volumetric flowrate	$[m^3/s]$
VOC	Variable operating cost	[EUR]
w	Width	[m]
W_s	Solid inventory of fluidized bed	[kg]
$W_{s,spec}$	Specific solid inventory of fluidized bed	$[kg/m^2]$
x	Mass concentration	$[wt.\%]$ $[-]$
X	Molar conversion	$[mol_{CaCO3}/mol_{Ca}]$
X_{ave}	Average molar conversion	$[mol_{CaCO3}/mol_{Ca}]$
$X_{max,ave}$	Maximum average sorbent activity	$[mol_{CaCO3}/mol_{Ca}]$
$X_{max,N}$	Maximum average sorbent activity having N complete carbonation-calcination cycles	$[mol_{CaCO3}/mol_{Ca}]$
X_r	Residual sorbent conversion capacity	$[mol_{CaCO3}/mol_{Ca}]$
y	Volumetric concentration	$[vol.\%][-]$

Greek symbols

δ_{CaL}	Share of additional net power by the CaL process	[–]
ε	Void fraction in a fluidized bed	[–]
ε_0	Initial particle porosity	[–]
ε_{mf}	Void fraction in a fluidized bed at minimum fluidization conditions	[–]
ε_{se}	Void fraction at the outlet of a riser	[–]
ε_s^*	Saturated solid carrying capacity of a gas	[–]
η	Efficiency	[–]
λ	Oxygen-to-fuel-ratio	[–]
μ	Dynamic viscosity	[Pa·s]
Λ	Specific make-up rate	[–]
ρ	Mass density	[kg/m³]
φ	Gas-solid contacting factor	[–]
Φ	Specific sorbent circulation rate	[–]
Φ_p	Particle porosity	[–]
τ_{active}	Active space time	[s]
Ω_{calc}	Calciner CO_2 partial pressure ratio	[–]

Subscripts and indices

0	Initial conditions, make-up limestone
abs	Absorbed
ave	Average
ar	As received (i.e. including ash and moisture)
bio	Biogenic carbon, biogenic CO_2
cal	Carbonate looping
$calc$	Calciner
$calcu$	Calculated
$capt$	Captured
$carb$	Carbonator
d	Dense region in fluidized bed
db	Dry basis (i.e. excluding moisture)
eq	Chemical equilibrium
ex	Existing system
f	Fluid
$flush$	Flushing gases
$foss$	Fossil
g	Gaseous phase
$gross$	Gross electrical efficiency
l	Lean region in fluidized bed
LS	Live steam
net	Net electrical efficiency
nom	Nominalized parameter

i	Inner
in	Inlet of an evaluation framework
o	Outer
out	Outlet of an evaluation framework
p	Particle
ref	Reference conditions
s	Solid state
$spec$	Specific
th	Thermal
tot	Total

Chemical symbols

Al_2O_3	Aluminum oxide
Ar	Argon
C	Carbon
C_2H_6	Ethane
C_3H_8	Propane
$C_6H_{12}O_6$	Glucose
Cl	Chlorine
CO	Carbon monoxide
CO_2	Carbon dioxide
Ca	Calcium
$CaCl_2$	Calcium chloride
CaO	Calcium oxide
$Ca(OH)_2$	Calcium hydroxide
$CaSO_4$	Calcium sulfate
$CaCO_3$	Calcium carbonate
CH_4	Methane
Fe_2O_3	Iron (III) oxide
Fe_3O_4	Iron (IV) oxide
H_2	Hydrogen
H_2O	Water
H_2S	Hydrogen sulphite
HCl	Hydrogen chlorine
HCN	Hydrocyanic acid
HFC	Hydrofluorocarbon
$MgCO_3$	Magnesium carbonate
N_2	Nitrogen
N_2O	Nitrous oxide
NH_3	Ammonia
NO	Nitrogen oxide
NO_2	Nitrogen dioxide

O_2	Oxygen
PFC	Perfluorocarbon
S	Sulfur
SF_6	Sulfur hexyfluorid
SiO_2	Silicon dioxide
SO_2	Sulfur dioxide
SO_3	Sulfur trioxide
SO_x	Sulfur oxides

Abbreviations

2CS	2 °C scenario proposed by the IEA
AF	Afforestation and reforestation
ASPEN	Commercial process simulation software
ASU	Air separation unit
B2C	Beyond 2 °C scenario proposed by the IEA
BA	Bottom ash
BECCS	Bioenergy carbon capture and storage
BFB	Bubbling fluidized bed
CA	Combustion air
CAD	Computer-aided design
CaL	Calcium looping, carbonate looping
CAPEX	Capital expenditure
CC	Combustion Chamber
CCS	Carbon capture and storage
CCU	Carbon capture and utilization
CFB	Circulating fluidized bed
CLC	Chemical looping combustion
COP	Conference of the parties
CSIC	Higher Council for Scientific Research
DAC	Direct air capture
DSI	Dry sorbent injection
ECO	Economizer
EF	Entrained flow
EGR	Enhanced gas recovery
EST	Institute for Energy Systems and Technology
EOR	Enhanced oil recovery
EU	European Union
EVA	Evaporator
EW	Enhanced weathering
FA	Fly ash
FB	Fluidized bed
FG	Flue gas
FOC	Fixed operational costs

FTIR	Fourier-transform-infrared-spectrometer
GHG	Greenhouse gas
GPU	Gas processing unit
HIC	Heat integration case
HP	High pressure section
IC	Indirect costs
IEA	International energy agency
IEAGHG	IEA greenhouse gas R&D program
IGCC	Integrated gasification combined cycle
INCAR	Institute of Carbon Science and Technology
IPCC	Intergovernmental panel on climate change
ITRI	Industrial technology research institute
LEP	Lignite energy pulverized
LHV	Lower heating value
LP	Low pressure
LPP	Low pressure preheater
LS	Loop seal
MB	Moving bed
MEA	Monoethanolamine
MSW	Municipal solid waste
MU	Make-up limestone
NDIR	Non-dispersive infrared
NET	Negative emission technology
NG	Natural gas
OC	Oxygen carrier, owner costs
OF	Ocean fertilization
OPEX	Operational expenditure
P1	Process contingency level
P2	Project contingency level
PA	Primary air
PSD	Particle size distribution
RDF	Residue derived fuel
RH	Reheater
SA	Secondary air
SH	Superheater
SRF	Solid recovered fuel
TDC	Total direct costs
TDCPC	Total direct costs including process contingency
TGA	Thermogravimetric analysis
TPC	Total plant costs
TUD	Technische Universität Darmstadt
TRL	Technology readiness level

VOC	Variable operating costs
WtE	Waste-to-energy
XRF	X-ray fluorescence

1 Introduction

1.1 Motivation

The continuous utilization of fossil energy resources and the related emissions of greenhouse gases (GHG) represents a major threat to humankind. The current scientific consensus is that the increase of GHG concentration in the atmosphere leads to a rise in the global-mean temperature on earth [1, 2]. Several challenging consequences such as a growing number of extreme weather events or the rise of seawater level are related to this phenomenon. Among the contribution of the GHG, such as methane (CH_4), water vapor (H_2O), nitrous oxide (N_2O), hydrofluorocarbons (HFCs), perfluorocarbons (PFCs) and sulfur hexafluoride (SF_6), carbon dioxide (CO_2) represents the major share [3].

As shown in Figure 1-1, there is a remarkable increase of the global average CO_2 concentration in the atmosphere, which is directly linked to the start of the industrial activities in the beginning of the 18th century. By the year 2017, a global average CO_2 concentration of more than 400 ppm$_v$ has been reached. This represents an increase of almost 50 % in comparison to the pre-industrial value of approximately 280 ppm$_v$. The rise of the CO_2 concentration is mainly caused by the extensive utilization of fossil-carbon based primary energy carriers, such as coal, oil or natural gas. These substances are typically combusted to generate heat, which is either directly used or further converted into other forms of energy, such as electricity. Nowadays, fossil fuels still represent the dominating primary energy carrier in almost all fields of relevant activities such as power generation, industry, transportation or building (i.e. heating) [4].

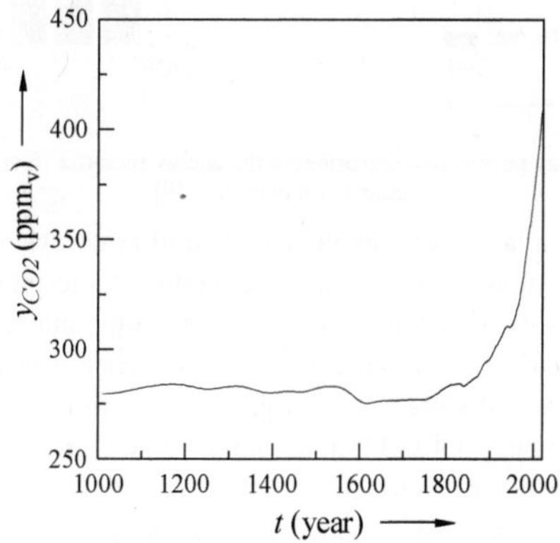

Figure 1-1: CO_2 concertation in the atmosphere from the year 1000 to 2017, data from [5, 6].

According to a data set from the International Energy Agency (IEA) [7], the global CO_2 emissions related to the combustion of fossil fuels accumulate to approximately 32.8 Gt$_{CO2}$ in 2017. This number represents the highest value ever noticed. The power and heat generation sector thus represents the major emitter with a total emitted amount of approximately 13.5 Gt$_{CO2}$/a, followed by the sectors of transportation (8.0 Gt_{CO2}/a) and industry (6.2 Gt_{CO2}/a). Amongst the latter, energy-intense industries such as steel- or clinker-making processes account for yearly CO_2 emissions of almost 1.7 Gt_{CO2}.

In order to limit the effects of climate change to an acceptable level, a drastic reduction of further GHG emissions is required. At the Paris Conference of the Parties (COP) 21 in 2015, the 195 nations agreed on the target to limit the increase of global-mean temperature to not more than 2 °C (preferably below 1.5 °C) in comparison to pre-industrial levels (1850 - 1900) [8]. For the energy sector, this implies changes in the way of energy supply, utilization and conversion. Additionally, new technologies need to be developed for an implementation in the long-term perspective.

In Figure 1-2, two energy technology perspectives proposed by the IEA from the years 2020 to 2060 are presented. Graph a) shows the "2 °C Scenario" (2CS), aiming on a CO_2 emission reduction, adequate for limit the rise of global-mean temperature at 2 °C compared to pre-industrial levels. Graph b) shows the "Beyond 2 °C Scenario" (B2CS), which aspires to a limit of the global-mean temperature rise at 1.5 °C compared to pre-industrial levels.

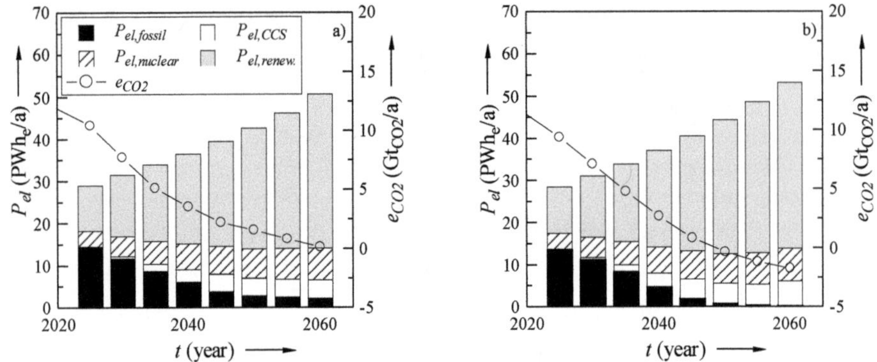

Figure 1-2: Energy technology perspective scenarios for the energy sector. 2 °C scenario (a) and beyond 2 °C scenario (b), data from [9].

Both scenarios are based on a large-scale implementation of renewable energy sources. In the year 2060, more than 70 % of the global power demand is supplied by technologies that rely on primary energy in the form of wind, solar radiation or hydropower. Furthermore, nuclear energy plays a key role for achieving the desired CO_2 emission reduction goal in each scenario. As a third measure, the application of carbon capture and storage (CCS) processes is foreseen for the second half of the century. In the 2CS, approximately 4,400 TWh$_e$ are supplied by CCS equipped power systems. In the B2CS, this number is forecasted to almost 5,800 TWh$_e$. Putting these numbers into perspective, it requires approximately 660 or 870 CCS equipped full-scale state-of-the-art power plants in 2060 for the 2CS and for the B2CS, respectively.

As a consequence of the proposed power generation infrastructure, the related CO_2 emissions decreases considerably. In the 2CS, the yearly CO_2 emissions decrease significantly in the first half of the century, whereas a moderate reduction is foreseen after 2045. This leads to CO_2 emissions due to the power generation sector of approximately 0.13 Gt_{CO2}/a in 2060. The reduction pathway is even more ambitious in the B2CS, where CO_2 emissions of -1.74 Gt_{CO2}/a are expected for the year 2060. Thus, CO_2 is removed from the atmosphere, while power is produced. Atmospheric CO_2 removal in the framework of power generation systems is feasible due to the utilization of biomass-based fuels in combination with CCS equipped power plants. Accordingly, the power generation in CCS equipped, biomass-fueled power plants accounts for approximately 2,214 TWh_e/a in the B2CS.

1.2   Objectives of the Thesis

Globally, there will be an increasing capacity of wastes that need to be handled in the future. According to the current technological knowledge, thermal waste treatment is seen as the most promising technology to be applied in this regard. As a consequence, Waste-to-Energy (WtE) plants will remain a relatively large CO_2 point emitter even in the long-term. In addition to that, there is the need to achieve negative CO_2 emissions in order to limit or even to reduce the CO_2 concentration in the atmosphere. Taking into account the organic waste fractions in Municipal solid waste (MSW), CCS equipped WtE plants could be applied to contribute to an achievement of both aforementioned needs. MSW is thereby treated in an environmentally friendly way, while heat and/or power, as well as negative CO_2 emissions, are supplied.

The carbonate or calcium (CaL) process represents an appropriate method for post-combustion CO_2 capture from newly built or already existing WtE plants. In this process, the capture of CO_2 from a flue gas stream is achieved by the cyclic carbonation-calcination of a circulating sorbent in two interconnected fluidized bed reactors, namely carbonator and calciner. The required heat to run the process is supplied by the combustion of supplementary fuels in the calciner. The feasibility of the process has already been experimentally demonstrated up to megawatt-scale using hard coal, lignite or natural gas as supplementary fuels. The CaL process is particularly suited for being retrofitted to existing plants. Among the processes for CO_2 that are currently researched, the CaL process shows a relatively low efficiency drop along with moderate CO_2 avoidance costs in the framework of a coal-fired power plant. Its application under the conditions of a WtE plant has neither experimentally nor techno-economically investigated yet. In such an arrangement, waste-derived fuels could be applied to heat the process. In contrast to the utilization of fossil-fuels, this implies low or even negative fuel costs, an improved carbon balance and a higher total waste throughput of the retrofitted WtE plant.

This thesis will classify the atmospheric CO_2 removal by means of CaL process-equipped WtE plants among other competitive technologies. A two-stage evaluation will be applied. First, an extensive experimental investigation of the CaL process under the boundary conditions given by the desired application is carried out at a 1 MW_{th} CaL pilot plant. This particularly implies for the first time worldwide the utilization of solid recovered fuel (SRF) in the calciner of the CaL process and the decarbonization of a flue gas similarly to that of a typical WtE plant fueled by MSW. Due to the strong influence of fuel properties on the CaL process performance, it is crucial to ensure the ability of the sorbent to absorb CO_2 by means of long-term investigations.

Based on a detailed experimental investigation, a comprehensive techno-economic assessment of the application of the CaL process for post-combustion CO_2 capture in the field of WtE plants is carried out. For the appropriate calculation of heat and mass balances, the CaL process model is validated by experimental data. The thermodynamic assessment covers a wide range of process conditions and possibilities for excess heat utilization. The economic calculations rely on a bottom-up approach for the investment costs of relevant components. In order to compare the techno-economic process performance among other technologies for atmospheric CO_2 removal, economical extreme scenarios are evaluated.

1.3 Thesis Outline

This thesis is structured as follows.

In Chapter 1, the challenges of modern society related to the extensive emission of GHG are described. Based on a general overview of climate change mechanisms, the need for negative CO_2 emissions is substantiated. Finally the objectives of this thesis and its structure are derived.

Chapter 2 provides the theoretical fundamentals that are associated with this thesis. This includes an overview of the global waste generation and the treatment of wastes in WtE plants. Thereafter, technical possibilities for climate change mitigation by means of CCS processes are provided. Each step of the CCS process chain is explained and discussed, in addition to the presentation of suitable processes that allow for negative CO_2 emissions. Based on the fundamentals of fluidization engineering and fluidized bed combustion, a state-of-the-art review of CaL process technology is presented. Particular attention is given to the carbonation-calcination reaction, sorbent deactivation mechanisms and existing experimental CaL process test facilities, as well as the techno-economic performance of the CaL process.

Chapter 3 focuses on the experimental investigations at the 1 MW_{th} CaL pilot plant. Based on a simplified process flowsheet, the main components and auxiliary systems of the CaL pilot plant at Technische Universität Darmstadt are explained. This includes the measurement devices for pressure, temperature, volumetric flow rate and gas composition as well as the auxiliary systems of the pilot plant. Thereafter, the main results of two consecutive CaL test campaigns at the pilot plant are presented and discussed. The evaluation is based on simplified reactor models for carbonator and calciner which are typically applied in the field of CaL technology. In addition to the CO_2 capture performance of the CaL process, a comprehensive evaluation of gaseous emissions related to the combustion of SRF is carried out. The specific emissions of CO, NO, SO_2 and HCl are thereby compared among different combustion conditions, with particular focus on the oxyfuel combustion under conditions of the CaL calciner.

In Chapter 4, a detailed process model for the calculation of the CO_2 absorption efficiency and sorbent conversion in the CaL carbonator is applied. Based on experimental data from the 1 MW_{th} pilot plant, the model is further validated. A sensitivity study points out effects of crucial process parameters on the CO_2 absorption characteristics in the carbonator. The validated CaL process is then applied to the techno-economic analysis in Chapter 5.

Chapter 5 is comprised of the techno-economic investigations of a generic 60 MW_{th} WtE plant being retrofitted by the SRF-fueled CaL process. Based on the heat and mass balances delivered by the validated process model, key performance indicators such as levelized cost of electricity or cost of CO_2 avoidance are calculated by a bottom-up approach. As the CaL technology is still at an early stage of development, a comprehensive sensitivity study is carried out in terms of technical and economical parameter variations. Additionally, novel concepts for CaL process excess heat integration into the existing WtE plant due to external super- and re-heating are proposed and thermodynamically evaluated.

Finally, Chapter 6 summarizes the main results of this thesis. Conclusions are drawn from the research and major fields for further investigations are proposed and highlighted.

2 Current State of Research

This chapter covers the current state of research in the fields that are relevant for this thesis. In particular, an overview of the current waste management strategies, the waste treatment in dedicated WtE plants and the waste processing and upgrading towards SRF is given. Thereafter, the theory of CCS is explained for each particular process step. Based on the fundamentals of fluidization engineering, the main mechanisms of the gaseous pollutant formation and mitigation with regard to the fluidized bed combustion of solid waste-derived fuels is explained. Finally, the current state of knowledge related to the CaL process is introduced.

2.1 Waste Management

In the course of the following section, an overview of global waste generation is presented. Different strategies are available for the treatments of waste streams. Among others, the thermal treatment in WtE plants represents one suitable option. The unique characteristics of WtE plants are then further introduced and described. Finally, the processes relevant for upgrading raw wastes towards more classified fuels is illustrated using SRF as an example.

2.1.1 Global Municipal Waste Generation and Treatment Strategies

The global generation of MSW is directly related to the growth rate of the population and the associated industrial activities. Its generation is consequently expected to grow from nearly 1.3 billion tons per year currently, to approximately 4.0 billion tons per year in 2025 [10]. In the EU-28, approximately 239 million tons of MSW were generated in 2014, which means 475 kg_{MSW} per capita, or 23 kg_{MSW} per 1,000 EUR of gross domestic product [11].

MSW consist of a wide range of substances. Exemplarily, Figure 2-1 depicts the mass composition of MSW for a low- and for a high-income level country, respectively. The mass compositions is classified towards organics (e.g. wood, food residues), papers (e.g. newspaper, cardboards), plastics (e.g. packaging, bags), glasses (e.g. bottles, colored glass), metals (cans, foil), textiles and others (leather, rubber).

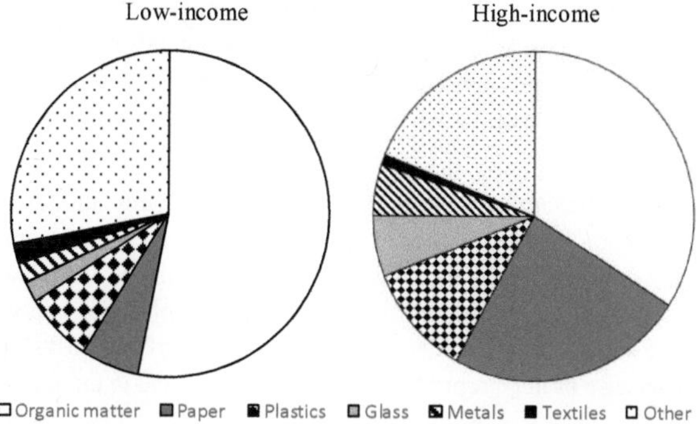

Figure 2-1: Composition of MSW dependent on the country income level, data from [12].

The composition of MSW strongly depends on the income level of the considered region/country. In a high-income country, the quantities of first-world waste products such as plastics, papers or glasses are higher in comparison to a low-income country. Contrary to that, the share of biogenic wastes is noticeable increased in regions with a low income level (34 % vs. 53 %).

Different methods are applied for the minimization and mitigation of wastes, which can be categorized according to the waste treatment hierarchy as shown in Figure 2-2. Accordingly, the approaches with the lowest environmental burden are classified at the top and should be preliminary applied. In the downward direction, the lifecycle impact of the waste treatment methods increase.

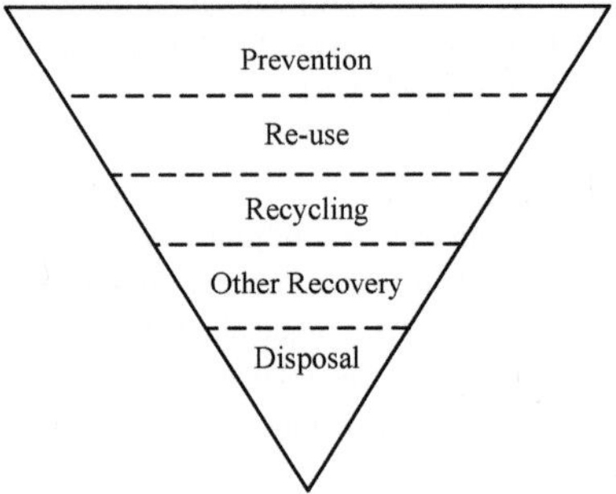

Figure 2-2: Waste treatment hierarchy, adopted from [13].

The prevention of waste represents the ideal approach, as neither resources are utilized to generate a certain product, nor does the product needs to be treated after its original purpose. Re-use is favorable to recycling, as the recycling step implies a process for the chemical or physical conversion of the old into the new product. One example is the direct re-use of packing material (e.g. bottle, box) instead of the recycling of the old plastic material to a new product. Other waste recovery approaches include thermal recovery by means of combustion processes or the chemical recovery by means of synthesis gas utilization via a gasification process step. The least desirable option is the disposal of wastes, as no further product or service is recovered from the waste substance after its desired utilization.

The currently applied strategies for the treatment of wastes that are neither applicable to recycling nor to re-use include the uncontrolled waste dumping, the landfilling of wastes with or without landfill gas collection and the incineration in WtE plants. In high-income countries, the thermal treatment of wastes or its controlled landfilling is mostly applied, while on the other hand low-income countries still dispose their wastes in uncontrolled dump sites [10]. In the EU, on an average basis 28 % of MSW are recycled, 27 % incinerated, 16 % composted and 28 % are landfilled [11].

In case of uncontrolled waste dumping, the emissions of methane as a consequence of the biological degradation processes significantly contribute to the global GHG emissions. According to a report from the U.S. Environmental Protection Agency, the methane emissions from uncontrolled landfills account for almost 12 % of global methane emissions [14]. Furthermore, the release and distribution of hazardous species such as heavy metals in the ecosystem is of major concern. Suitable waste treatment strategies are thus required in order to limit the pollution of the ecosystem and the emission of greenhouse gases because of inappropriate waste disposal now and in the future.

2.1.2 Solid Recovered Fuel

The term solid recovered fuel refers to a specially pre-treated type of waste-derived fuel, processed from production-specific wastes, non-hazardous industrial wastes or MSW [15]. It is the successor of the refuse derived fuel (RDF), which has come onto the market during the 1970s.

The main constitutes of SRF are plastics, paper, cardboard, textiles and wood. Plastics could be in the form of thin foils or hard plastic structures in various 3D shapes. Wood might be derived from demolition wood or production residues. The major steps within an SRF processing plant include shredding, metal separation, sieving and size reduction. SRF is typically utilized in dedicated mono-combustion units, cement plants and as a substitute fuel in fossil-fired power and industrial plants [16-18]. The different fuel constitutes have a considerable influence on the combustion properties of SRF. As the composition of SRF is highly dependent on countries, regions and seasons, a classification approach has been proposed by the EU [19]. The following key classification characteristics were introduced:

- Mean net calorific value, in MJ_{th}/kg (as received)

- Mean chlorine content, wt.% (dry)

- Median and 80th percentile mercury content, mg/MJ_{th} (as received)

Based on the above mentioned characteristics, five different SRF quality classes are distinguished according to Table 2-1. The diverse nature of SRF becomes apparent when taking into account the different net calorific values. While the net calorific value of Class 1 is comparable to the corresponding number of a typical hard coal, the energy content of Class 5 is almost insignificant.

Table 2-1: Classification characteristics and quality classes of solid recovered fuel [20].

Characteristic	Statistical measure	Classes				
		1	2	3	4	5
Net calorific value	Mean	≥ 25.0	≥ 20.0	≥ 15.0	≥ 10.0	≥ 3.00
Chlorine	Mean	≤ 0.20	≤ 0.60	≤ 1.00	≤ 1.50	≤ 3.00
Mercury	Median	≤ 0.02	≤ 0.03	≤ 0.08	≤ 0.15	≤ 0.50
	80th percentile	≤ 0.04	≤ 0.06	≤ 0.16	≤ 0.30	≤ 1.00

In contrast to traditional fossil fuels such as coal, oil or natural gas, SRF offer the advantage of a moderate carbon footprint due to organic waste fractions. Furthermore, fuel prices are usually notably reduced in comparison to conventional fossil fuels. Depending on the regulatory framework and the waste-management system, even negative fuel prices are feasible [21]. In contrast, the requirements of the combustion system increase due to challenging fuel species contained in SRF. In this respect, chlorine poses the main threat because of the risk of high-temperature corrosion. In order to prevent boiler materials from exhausting corrosion rates, the temperature of the live steam is typically reduced (see Chapter 2.1.2). Another challenging aspect of SRF is the presence of coarse mineral matter and incombustible substances. Therefore, special attention needs to be paid to the design of the ash extraction systems of SRF-fueled combustion units.

Figure 2-3 gives a comparison of the specific CO_2 emissions (e_{CO2}) for different types of fuels referring to the total and fossil share of CO_2, respectively. For biomass a biogenic carbon fraction of 100 % is assumed, with 40, 45 and 65 % for SRF I, SRF II and MSW. In the case of the latter three fuel types, a variation in the biogenic carbon fraction of +/- 20 % is additionally indicated.

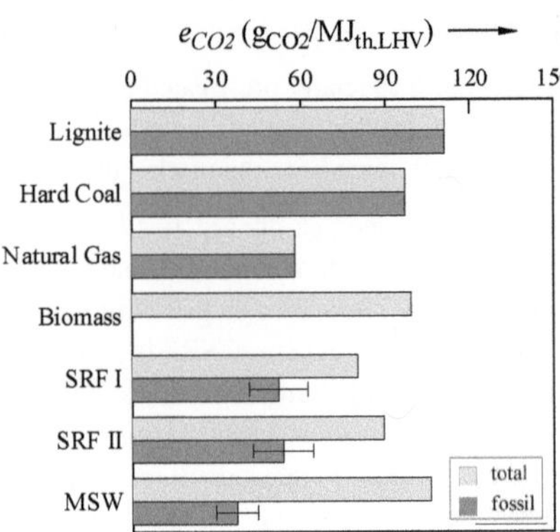

Figure 2-3: Comparison of the specific CO_2 emissions (e_{CO2}) for different types of fuel, data from [22-24].

The total specific CO_2 emissions are least favorable in case of lignite (e_{CO2}: 111 g$_{CO2}$/MJ$_{th,LHV}$) and MSW (e_{CO2}: 106 g$_{CO2}$/MJ$_{th,LHV}$). On the other hand, natural gas allows for the lowest total specific CO_2 emissions (e_{CO2}: 57.8 g$_{CO2}$/MJ$_{th,LHV}$). If only the share of fossil CO_2 is considered, the comparison shows a different characteristic. The utilization of pure biomass is considered as almost CO_2 neutral, which results in net zero CO_2 emissions. The second best option in this regard is MSW; its combustion leads to specific CO_2 emissions of 37.1 g$_{CO2}$/MJ$_{th,LHV}$, followed by the two types of SRF with approximately 52.8 g$_{CO2}$/MJ$_{th,LHV}$. Even though the large-scale combustion of pure biomass might not be fully harmonized with the planetary boundaries, the utilization of waste-derived fuels allows for a better specific CO_2 emissions balance than all other fossil fuels typically considered for power generation. Due to increasing share of biogenic feedstock in the chemical industry that is forecasted in the long-term, it is likely that the fossil-fraction in waste-derived fuels will decrease even further in the future.

2.1.3 Waste-to-Energy Plants

The incineration of raw MSW in WtE plants is a widely used waste treatment methodology. The main purpose of a WtE plant is the reduction of waste mass, volume and toxicity while generating power and/or heat by the utilization of the energy chemically bound in the MSW as a useful byproduct [25].These days, nearly 28 % of the MSW capacity in the EU-28 is treated by means of incineration [26]. Worldwide, more than 750 WtE plants with a yearly MSW treatment capacity of almost 83 million tons are currently in operation [12].

The net electrical efficiency of a typical WtE plant ranges from 10 to 30 %. This relatively low value is caused by several reasons, such as the low energy content of the raw MSW, its fluctuation in composition and size as well as the various corrosive species contained in the MSW reducing the effectiveness of heat recovery. Combustion systems for MSW, such as moving grates or fluidized beds, are specifically designed to guarantee a stable plant operation, complete fuel burnout along with non-toxic emissions rather than aiming at maximum boiler efficiency. The power cycle of a WtE plant is exposed to multiple material demanding phenomena, such as high temperature corrosion, molten salt corrosion, fouling and the erosion of heat exchanger surfaces [27, 28]. In addition to the high temperature corrosion on boiler and superheater surfaces, there is also the risk of low-temperature corrosion [29]. To prevent the low-temperature corrosion at the outlet of boiler and in the flue gas treatment path, the minimum temperature from exiting flue gas streams is limited, which again lowers the water-steam cycle efficiency. Furthermore, WtE plants are often applied for the delivery of heat as the main product, which raises the need for the application of the total energy efficiency rather than only taking into account the net electrical efficiency. Exemplarily, Figure 2-4 shows the energy flows of WtE plants delivering power only (a) and heat and power (b).

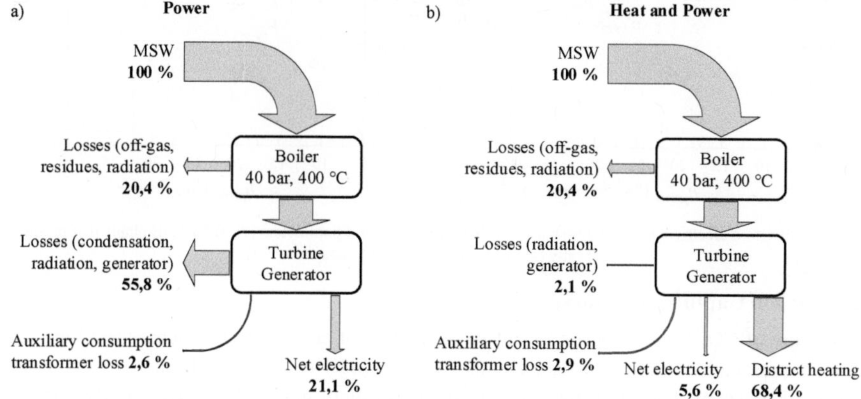

Figure 2-4: Energy flows in WtE plants that produce power (a) or heat and power (b), adopted from [30].

The parameters of the superheated steam at the turbine inlet are similar in both plants, while the total energy output differs significantly between both approaches. When producing heat and power, 74.4 % of the chemical energy in the MSW feed is converted into the desired product, whereas this number is reduced to 21.2 % in case of power as the only product. However, electricity output in the heat and power case is limited to 5.6 % of total energy input, while 68.4 % of total energy input is converted to hot water for district heating applications. Energy losses related to the condensation of steam at the turbine outlet are thus minimized in the heat and power case.

A WtE plant represents a highly case specific system that is tailored to fit the prevailing boundary conditions, such as local stack emission limits, fuel composition and the desired output product [31]. Especially the water-steam cycle layout varies among different applications. For example, Figure 2-5 shows the live steam parameters as a function of the gross electrical efficiency for WtE plants located in the EU.

In order to reduce the high temperature corrosion induced material losses to an acceptable value, the temperature of the superheated steam is often limited to approximately 400 °C [32]. For both types of WtE plants, the major share of the units thus operate at live steam parameters of 400 °C and 40 bar, leading to a gross electrical efficiency between 20 % and 30 %. As the extent of corrosion induced material losses is related to the chemical composition of the fuel and to the mitigation measure that are applicable under certain conditions, some of the WtE plants operate with a live steam temperature even above 500 °C and 75 bar. This results in an increased gross electrical efficiency of 30 - 35 %. Furthermore, the more diverse nature of heat and power producing WtE plants in terms of water-steam cycle layout becomes apparent when comparing both graphs.

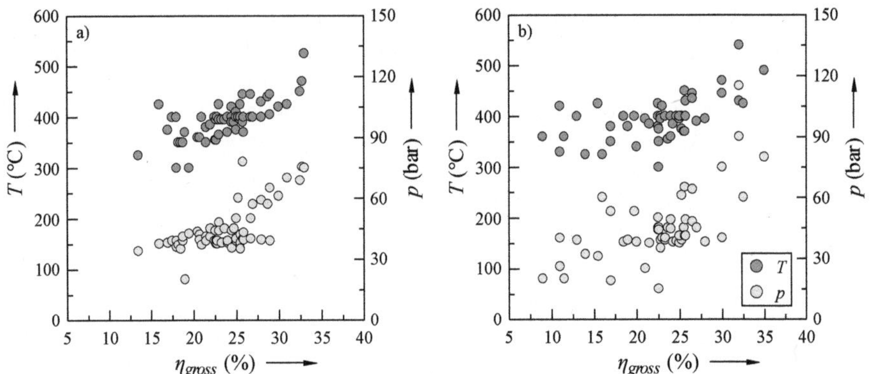

Figure 2-5: Gross electrical efficiencies (η_{gross}) for different live steam parameters (T, p) for WtE plants that produce power only (a) or heat and power (b), data from [33].

2.2   Carbon Capture and Storage

The capture of CO_2 from exhaust gases of large-scale point emitters and its subsequent long-term storage represents one option to reduce anthropogenic GHG emissions [34, 35]. A CCS system consists of processes for the capture of CO_2, its purification, transportation and injection into an appropriate long-term storage reservoir.

Furthermore, the highly concentrated CO_2 is suitable to be utilized as raw material for chemical processes or synthetic fuel [36, 37]. Consequently, the process chain is called carbon capture and utilization (CCU). In the case of a CCU application, the long-term CO_2 avoidance potential highly depends on the type of product that is derived and from the quantity of fossil feedstock that is replaced by the captured carbon. For example, the conversion of CO_2 into a synthetic fuel, such as methane or methanol, inevitably leads to the re-release of CO_2 once the fuel is combusted. Thus, it is rather complex to determine the positive effect of CCU with regard to the long-term CO_2 emission mitigation potential. The framework of this study is limited to the assessment of CCS and the related long-term storage of CO_2.

2.2.1   Capture

Within a CCS process chain, the process step of CO_2 capture represents the most energy and capital intense process step. Different approaches for CO_2 capture are distinguished according to the occurrence of the CO_2 capture process relatively to the stage of CO_2 formation (i.e. combustion). Figure 2-6 shows these three approaches for CO_2 capture applied to a process for the delivery of heat and/or power. The type of fuel might differ among coal, biomass, natural gas and waste or waste-derived fuels. Their common factor is the presence of carbon in the fuel which causes the formation and subsequent emission of CO_2.

The pre-combustion approach involves the capture of CO_2 upstream of the combustion process. In the case of a solid fuel, the first process step comprises the gasification of the fuel in order to produce a syngas that contains mainly H_2, CO and CO_2. For gaseous fuels, the syngas is achieved by means of a reforming process. Thereafter, the syngas undergoes several cleaning and upgrading steps, such as CO-shift and CO_2 capture in order to achieve an almost pure H_2-stream that is subsequently oxidized in a dedicated gas-turbine. The CO_2 capture from the syngas is typically carried out via Selexol or Rectisol processes [38]. The pre-combustion CO_2 capture concept has been realized using the Integrated Gasification Combined Cycle (IGCC) process [39, 40].

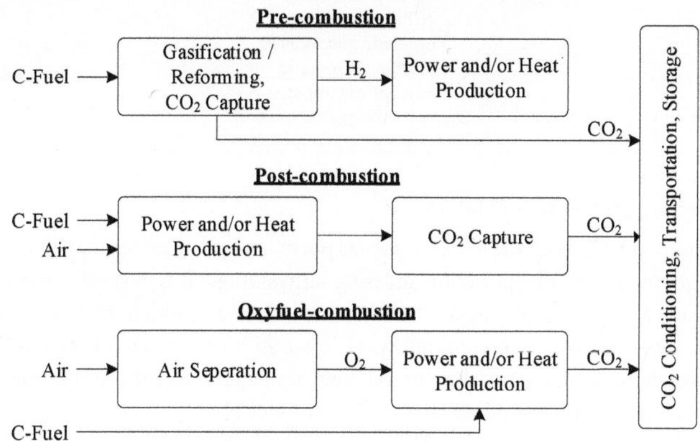

Figure 2-6: Approaches to capture CO_2 from combustion processes, adopted from [41].

The post-combustion approach involves the capture of CO_2 downstream of the combustion process. The capture of CO_2 thus represents an additional step in the flue gas cleaning line of a power plant. Consequently, post-combustion CO_2 capture processes are particularly suited for being retrofitted to already existing systems. Exemplary processes for post-combustion CO_2 capture are based on chemical or physical absorption. The chemical absorption by aqueous monoethanolamine (MEA) solvents is already employed at large-scale at the Boundary Dam power station in Canada [42, 43]. Other processes such as the chilled ammonia process [44, 45] or the CO_2 capture by means of membranes are currently being investigated [46, 47]. The CaL process is also categorized as post-combustion CO_2 capture process (see Chapter 2.3).

Oxyfuel combustion refers to fuel oxidation in the absence of air-nitrogen. The resulting flue gas mainly consists of CO_2 and H_2O. After condensation of the water vapor, there is an almost pure CO_2 stream available. The air-oxygen is replaced by technically pure oxygen. In order to avoid thermal induced agglomeration and sintering phenomena in the combustion zone, part of the flue gas is recirculated for the dilution of oxygen. In the most mature oxyfuel combustion process layout, the oxygen in the oxidation agent is provided directly by an on-site cryogenic air separation unit (ASU) [48-50]. In order to avoid the efficiency penalties related to the operation of the ASU, the indirect supply of the oxygen by more sophisticated process approaches, such as chemical looping, is currently being researched. In this process, a solid oxygen carrier (OC) circulates between air and fuel reactor. In the fuel reactor, the OC releases oxygen for fuel combustion. After being transferred to the air reactor, the OC is oxidized again [51, 52].

Table 2-2 summarizes the most common CO_2 capture processes within each approach. Additionally, the technology readiness level (TRL) is given for each option.

Table 2-2: CO_2 capture approaches and the current status of process development, data from [53-55].

CO_2 capture approach	Process	TRL[1]
Pre-combustion	IGCC	7
Post-combustion	Amine scrubbing	9
	Chilled ammonia	7
	Polymeric membranes	6
	Carbonate looping	6
Oxyfuel-combustion	Oxyfuel gas turbines	2-5
	Oxyfuel combustion of coal	7
	Chemical looping combustion	6

2.2.2 Conditioning and Transportation

After the separation of CO_2 from the exhaust gas, its purity and pressure need to be increased in order to fulfill the requirements for transportation and long-term storage. It is desired to transfer the gaseous CO_2 into the supercritical state above 31.0 °C and 73.8 bar, which makes the long-distance transportation reasonable. The maximum allowable quantities of species other than CO_2 depend on the type of transport systems (e.g. ship or pipeline transport) and the limitation related to the considered storage side. The upgrading of the CO_2 product is carried out in a gas processing unit (GPU). A GPU comprises process steps such as compression, intercooling and condensation in order to deliver the CO_2 in the purified, supercritical state [56].

The CO_2 transport via pipelines is already established in the course of enhanced oil recovery (EOR) projects in the United Sates. Table 2-3 exemplarily summarizes the purification requirements for the Weyburn EOR project and those defined in the course of the Dynamis research project. The primary reasons for the limitation of minority species in the CO_2 product stream are associated with the risk of corrosion in the pipeline, health and safety issues and the geophysics of supercritical CO_2 in the storage site [57].

Table 2-3: Quality specification for the pipeline transport of CO_2 [58].

Component	Weyburn EOR [35]	Dynamis [59]
CO_2	> 96 vol.%	> 95.5 vol.%
Ar	-	< 4 vol.%
C_xH_y	< 0.7	< 4 vol.% (saline aquifers) < 2 vol.% (EOR)
CO	< 1000 ppm$_v$	< 2000 ppm$_v$
H_2O	< 20 ppm$_v$	< 500 ppm$_v$
H_2S	< 9000 ppm$_v$	< 200 ppm$_v$
N_2	< 300 ppm$_v$	< 4 vol.%
NO_x	-	< 100 ppm$_v$
O_2	< 50 ppm$_v$	< 4 vol.% (saline aquifers) < 100 - 1000 ppm$_v$ (EOR)
SO_x	-	-
Temperature	< 50 °C	-

The quality specification for the CO_2 product is of concern as the power demand and the costs related to the operation of the GPU are mainly driven by the upgrading requirements. In the same vein, the

---

[1] TRL 1: Basic principle established, TRL 2: Technology concept formulated, TRL 3: Experimental proof of concept, TRL 4: Technology validated in lab scale (continuous operation P_{th} < 50 kW$_{th}$), TRL 5: Technology validated in relevant environment (50 kW$_{th}$ < P_{th} < 1,000 kW$_{th}$), TRL 6: Technology demonstrated in industrial environment (P_{th} > 1,000 kW$_{th}$), TRL 7: System prototype demonstration (P_{th} > 10 MW$_{th}$), TRL 8: System complete and qualified (industrial scale), TRL 9: System proven in operational environment (competitive manufacturing of full system)

type of CO_2 capture process is also to be taken into account. For example, the upgrading of the CO_2 product stream from a MEA scrubbing process is by far less energy-intense than the upgrading of an oxyfuel combustion derived CO_2 product stream.

2.2.3 Storage

To ensure the CO_2 mitigation by CCS, a safe and controlled long-term storage of CO_2 is required. Currently, geological storage seems the most likely solution. The CO_2 is introduced into suitable deep rock formations, namely saline aquifers, natural gas and oil fields as well as undevelopable coal deposits.

Saline aquifers offers the largest CO_2 storage potential in the world [60]. In principle, these geological formations are similar to those of natural gas and petroleum reservoirs. In contrast, salt water is found in the pores of the rock layers instead of the aforementioned fossil energy carriers. However, before CO_2 sequestration, saline aquifers still have to be geologically investigated to ensure their suitability. The Sleipner Field is the world's first large-scale CO_2 injection project which is based on a saline aquifer structure. Since 1996, CO_2 from natural gas purification has been injected into the 1,000 m depth Utsira Sand formation under the North Sea [61].

Oil and natural gas fields have already proven to be secure storage options, so they seem safe as long-term deposits. Furthermore, the dimensions and the seismic formation structures are already known due to the geological investigations related to the exploitation. If the CO_2 is injected into an active oil or gas field, the pressure of the reservoir increases, which enhances the yield of reservoir. Such a procedure is called EOR or enhanced gas recovery (EGR) in case of a natural gas field. At the world's first commercial post-combustion CSC project at the Boundary Dam power station in Canada, more than one million tons of CO_2 per year are captured from a 139 MW_e coal-fired power station. The captured CO_2 is subsequently used for EOR in the Weyburn oil field [62].

Due to the pore structure of the coal, it adsorbs CO_2. Coal deposits at depths of 1,500 m or more are not economically viable according to current knowledge and are therefore suitable for CO_2 storage. The CO_2 is injected into the coal seam and deposits on the surface structure of the coal. As the greenhouse gas is firmly bound to the coal, it cannot volatilize in the long-term [63].

2.2.4 Negative CO_2 Emissions

According to the energy technology perspectives proposed by the IEA [9, 64], there will be the need to remove CO_2 from the atmosphere by the end of this century. Removal of atmospheric CO_2 is feasible by the implementation of negative emission technologies (NET). Figure 2-7 summarizes the major classes of NETs that are currently being discussed and their interaction with the major carbon reservoirs on earth, i.e. atmosphere, land, ocean and geological formations. Bioenergy with CCS (BECCS) and direct air capture (DAC) systems are technical approaches, whereas afforestation and reforestation (AF) and ocean fertilization (OF) are considered bioengineering methods. The enhanced weathering (EW) is seen as a geoengineering technology.

BECCS is associated with the utilization of bioenergy in a power or industrial plant in combination with a CCS system [65]. Thereby, CO_2 that has been formed during the combustion of biomass is captured and subsequently stored. Taking into account the carbon that has been bound during the growth of the biomass via the formation of glucose ($C_6H_{12}O_6$) and O_2 (i.e. photosynthesis, Eq. 2-1), CO_2 is removed from the atmosphere. The combustion of biomass might be carried out in mono-combustion systems or by means of co-combustion of biomass in already existing fossil-fueled units. Due to the similarities in the fundamentals of fossil and bioenergy CCS system, BECCS represents a relatively cost efficient and well-developed NET.

$$6H_2O + 6CO_2 \rightarrow C_6H_{12}O_6 + 6O_2 \qquad (2\text{-}1)$$

As considerable amounts of biomass are required in order to achieve sufficient quantities of CO_2 removal by means of BECCS, the sustainability of the bioenergy feedstock is of major concern; in particularly the competition for arable land and fresh water between energy and nutrition crops [66, 67]. A more sustainable attempt in this vein is the utilization of waste biomasses such as MSW or waste-derived fuels as a resource for bioenergy in BECCS applications [68, 69].

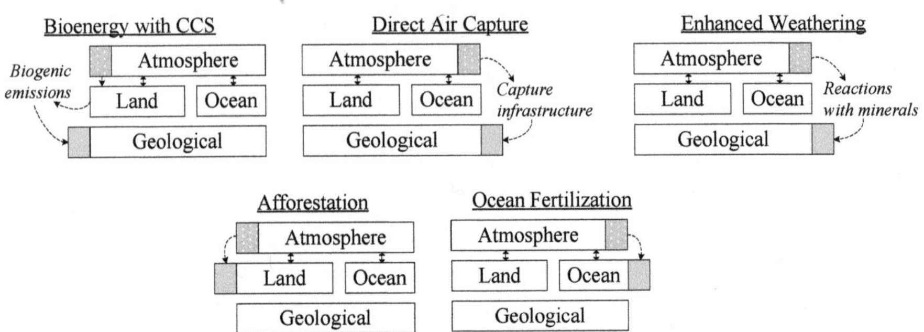

Figure 2-7: Schematic of NETs including the representation of carbon flows, adopted from [70].

The CO_2 capture directly from ambient air is referred to as DAC. The DAC CO_2 capture process relies on technologies such as absorption or adsorption [71, 72]. Due to the very low CO_2 concentration in the ambient air (approx. 410 ppm_v), DAC technologies are prone to a large specific energy consumption per mass of CO_2 that is captured [73]. By DAC applications, the CO_2 emissions from dispersed sources (e.g. transportation, aviation sector) are captured as well in addition to the CO_2 emissions from concentrated sources (e.g. power plants- or industrial plants). As ambient air essentially serves as a means of transportation for CO_2, the DAC unit could be placed independently of the location of the CO_2 emitter.

The absorption of atmospheric CO_2 by mineral matter is a continuously occurring process in nature. EW as geoengineering technology describes the enhancement of this process by chemical or physical means, such as the enhanced weathering of rocks. This is achieved by crushing suitable rocks to small particles that allow for a high specific surface area, and spreading them in an appropriate climate that enhances the naturally occurring CO_2 absorption [74]. Other approaches rely on the utilization of diluted solutions, which results in considerable water requirements [75].

AF is a well-understood and already applied method. Avoiding the excessive deforestation of forests and the planting of new trees increases the carbon reservoir on land. In addition of being a huge CO_2 sink, forests provide other services such as freshwater supply, wildlife habitat or preserving the biodiversity [76]. OF describes the approach of increasing the biomass productivity in the ocean. As microalgae have a high photosynthesis efficiency, large quantities of CO_2 are convertible into organic carbon. Potential biological substances include microalgae and cyanobacteria [77].

2.2.5   Carbon Capture and Storage in the Framework of Waste-to-Energy Plants

The application of CCS in the framework of WtE plants has rarely been discussed. This is partially due to the fact that the total amount of CO_2 that is globally emitted by WtE plants has been relatively low until now. Furthermore, in comparison to typical processes that are considered for CCS, such as large-scale power or industrial plants, WtE plants show some disadvantages which have a negative impact on the techno-economic process performance of a CCS retrofitted WtE plant. This is for instance the relatively poor net electrical efficiency, which causes a large quantity of CO_2 that needs to be captured per quantity of service (e.g. heat and/or power) that is provided. Furthermore, the CO_2 volumetric concentration in the flue gas is approximately 10 vol. %, which makes the process of CO_2 capture more energy intense compared to other industries that emit CO_2 in higher concentrations. With regard to process economics, the relatively small plant size increases the specific investment costs of a CCS system.

Recently, more research of CCS in the framework of WtE plants has been done due to several reasons. The rapid growth of global population and waste generation leads to a growing demand for suitable waste treatment strategies (see Chapter 2.1.1). Until now, uncontrolled landfilling and dumping are widely applied as a waste treatment method. Considerable amounts of GHG (e.g. CO_2, CH_4) are emitted by the biological degradation processes of wastes. WtE plants were identified to play a key role for an environmentally friendly waste treatment in the future. While heat and/or power are generated by the incineration of wastes, the toxicity and mass of the waste streams are efficiently reduced. Furthermore, WtE plants have been identified to represent a suitable host process for the supply of negative CO_2 emissions, in case of a CCS retrofit [78]. In a recent study, it was found that approximately 2.8 billion tons of negative CO_2 emissions per year are feasible by 2100, if all the available MSW is treated in CCS equipped WtE plants [68]. For the sake of comparison, it needs to be noted that the global energy-related CO_2 emissions according to the IEA B2C scenario need to be as low as -1.74 Gt_{CO2}/a (see Figure 2-2). Thus on global scale, there is a relevant potential of using MSW as a resource for negative CO_2 emissions.

For the Klemetsrud combined heat and power WtE plant in Oslo (Norway), a feasibility study for CO_2 capture by an aqueous solvent shows the potential capturing of 90 % of the total CO_2 emitted by the plant [79]. This means a yearly quantity of 400.000 tons of CO_2 which are subsequently transported via ship to an off-shore storage side. At the Twence WtE plant in Hengelo (Netherlands), approximately 830.000 tons of wastes are incinerated per year, while electricity and heat are supplied. Aker solution will deliver a CO_2 capture system suitable to capture 10.000 - 100.000 tons per year based on their "Just Catch" technology [80]. The operation is expected to start in 2021. At the WtE plant in Saga City (Japan), a Toshiba-designed CO_2 capture facility based on a solvent-scrubbing process has been capturing approximately 10 tons per day from the flue gas stream since 2016. The CO_2 is subsequently utilized for agriculture purposes [81].

The application of the CaL process for CO_2 capture from WtE plant flue gases was first considered in the work of Kremer in 2014 [82]. Here, the combustion of raw MSW or coal in the calciner has been assessed by means of a thermodynamic process model. It was already found that the external superheating of WtE live steam by CaL process excess heat might increase the overall net electrical efficiency of the retrofitted system. However, this study does neither investigate any process economics, nor does it experimentally prove the utilization of MSW in a CaL calciner.

2.3   Fundamentals of Fluidized Beds

In a fluidized bed, solid particles are being transformed into a liquid-like fluidized state. This is achieved by the upwards flow of a gaseous or a liquid stream around bulk solid particles. Fluidized beds have been originally developed for the coal gasification application by Winkler in the early 1920s. Until now, fluidized beds had been widely used in different applications such as, synthesis reactions, cracking of hydrocarbons or combustion, gasification and calcination processes. Fluidized bed systems are characterized by the following advantages [83]:

- The liquid-like flow of particles allows continuous operation.

- Nearly isothermal conditions in the reactor due to rapid mixing of particles.

- The whole solid reactor inventory represents a large safety margin to avoid temperature peaks.

- Good conditions for heat and mass transfer.

On the contrary, the application of fluidized beds poses the follow challenges [83]:

- Erosion of pipes and vessels due to abrasion induced by the particles.

- Risk of agglomeration and sintering of fine particles at high temperatures.

- Non-uniform residence time of particles in continuous operation.

2.3.1   Particle Characterization

The fluidization behavior of particles within a gas/solid fluidized bed is influenced by the properties of the particles. The particle mass density and the particle diameter are the governing particle characterization measures in this way. According to Geldart [84], particles are classified in four distinct groups by their diameter and the difference of particle and gas mass density.

Figure 2-8 shows the particle classification of the four groups (C: "cohesive", A: "aeratable", B: "sand-like", D: "spoutable") based on an experiment with ambient air as fluidization agent.

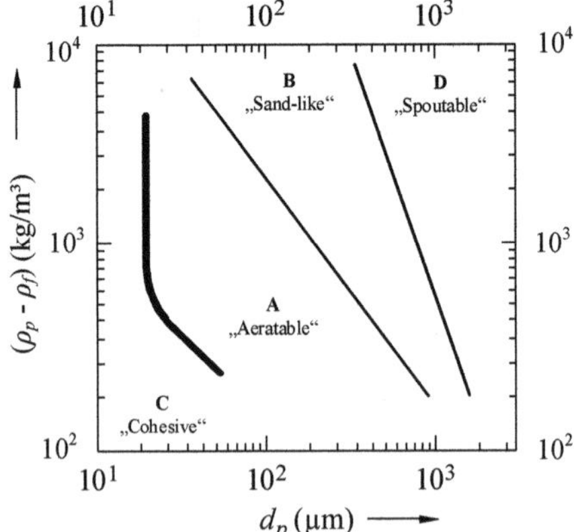

Figure 2-8: Geldart particle classification at ambient conditions; adopted from [83].

Group "C" particles: Particles or powders of this group are relatively small (particle size: < 30 μm) and strongly cohesive. This can be attributed to the fact that the interparticle forces are greater than the forces induced from the fluidization agent. A fluidized bed of group "C" particles tends to form channels, where part of the gas flow passes through the bed without fluidizing the particles. A proper fluidization behavior is therefore difficult to achieve. In order to process group "C" powders, they might be mixed with coarser particles of the same class.

Group "A" particles: "Aeretable" particles have a relatively small particle diameter along with a low mass density. They can be fluidized easily and smoothly. The formation of bubbles in the bed appears only at a considerable bed expansion after the gas velocity surpasses the minimum fluidization velocity. The gas bubbles reach a maximum diameter of approximately 10 cm and rise faster than the gas passes through the suspension phase. Once the fluidization stops, the bed collapses slowly with a velocity similar to the superficial velocity of the suspension phase.

Group "B" particles: The diameter of group "B" particles range from 40 μm - 500 μm with a typical mass density of 1,400 kg/m³ up to 4,000 kg/m³. Particles of this group are also called "sand-like" particles. The fluidization behavior differs significantly compared to group "A" particles. The bubble formation starts immediately once the superficial gas velocity surpasses the minimal gas velocity. The maximum bubble size is not limited by gas/solid fluidization interactions and can reach relatively large dimensions. The greatest share of the bubbles rise faster than the gas in the suspension phase. The bed expansion is small, and the bed collapses quickly, in case of an interruption of fluidization.

Group "D" particles: Particles of this group are the largest in terms of diameter and mass density. Particle diameter is larger than 600 μm. Fluidization is thus difficult to achieve, and a large flow of fluidization gas is needed. In some cases, even more than that is required for the chemical or physical process within the fluidized bed. Fluidization characteristics are spouted-like. Bubble formation is similar to group "B" particles. However, the bubbles rise with a lower velocity than the gas within the suspension phase.

2.3.2 Fluidization Regimes

The phenomenon of fluidization starts once the drag force of the fluidization agent (e.g. gas or liquid) equals or exceeds the weight force of the particle. Here, only the case of a gaseous fluidization agent is taken into account. Depending on the superficial gas velocity of the fluidization gas and the particle properties of the bulk solid phase, different fluidization regimes could be distinguished. The superficial gas velocity, u_0, in a reactor riser is calculated according to Eq. 2-2, where, $\dot{V}_{gas}$ is the volumetric flow rate of the gas, and A is the cross-sectional area of the reactor riser:

$$u_0 = \frac{\dot{V}_{gas}}{A} \tag{2-2}$$

In addition to the superficial gas velocity, the void fraction of the bed, ε, is typically applied for characterization of a fluidization regime:

$$\varepsilon = \frac{V_{tot} - V_s}{V_{tot}} \tag{2-3}$$

Here, V_{tot} is the total volume of the fluidized bed reactor and, V_s is the volume occupied by the particles herein. Figure 2-9 summarizes the characteristic fluidization regimes and the corresponding qualitative pressure drop depending on the superficial gas velocity.

Due to the fact that the solid phase usually shows a distribution of the particle size, a combination of different fluidization regimes appear at the same time. This means, for example, that at the transition point from a packed to a fluidized bed, small particles are already fluidized, whereas the big particles still remain in the packed stage.

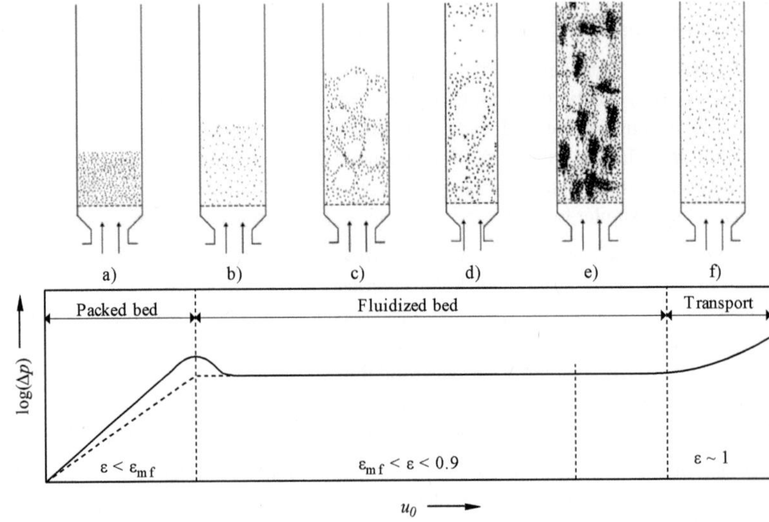

Figure 2-9: Fluidization states dependent on the superficial gas velocity, adopted from [85].

At relatively low fluidization velocities, the particles remain as a packed bed, and the fluidization agent passes through the interparticle spaces as the drag force of the gas flow is insufficient to move the particles (Figure 2-9a). In the stage of a packed bed, the bed height, h, does not change, whereas the pressure drop, Δp increases according to the semi-empirical Ergun-Equation (Eq. 2-4) as a function of the superficial gas velocity [85] for polydisperse systems:

$$\frac{\Delta p}{h} = 150 \frac{(1 - \varepsilon)^2}{\varepsilon^3} \frac{\mu u_0}{d_p{}^2} + 1.75 \frac{(1 - \varepsilon)}{\varepsilon^3} \frac{\rho_g u_0^2}{d_p{}^2} \tag{2-4}$$

Here, μ is the dynamic viscosity of the gas, d_p is the diameter of the particels and, ρ_g is the mass density of the gas.

Once the superficial gas velocity is raised to the minimum fluidization velocity, u_{mf}, the movement of the particles starts and the bed height expands (Figure 2-9b). The minimum fluidization velocity can be derived by the following equation:

$$u_{mf} = \frac{d_p^2(\rho_p - \rho_f)g}{150\mu} \frac{\varepsilon_{mf}^3 \phi_p^2}{1 - \varepsilon_{mf}} \tag{2-5}$$

Here, ϕ_p is the particle porosity.

The pressure drop of the fluidized bed at this stage is calculated as follows:

$$\Delta p = h(1 - \varepsilon)(\rho_g - \rho_f)g \tag{2-6}$$

At superficial gas velocities above the u_{mf}, the formation of bubbles inside the bed starts. Here, the gas flow moves up in particle-free bubbles. This fluidization stage is referred to as bubbling bed (Figure 2-9c).

As the size of the bubbles expands during the upwards movement in a riser, the bubble size could reach the full cross sectional area of a riser. In such a case, the regime is called slagging or spouting bed (Figure 2-9d). This phenomenon is likely to occur in risers that show a high height/diameter ratio. If the superficial gas velocity exceeds the terminal velocity of the particles, the particles are entrained from the riser. The terminal velocity of particles refers to the velocity in free-fall conditions at which the particles are not further accelerated. Consequently, particles are distributed over the total height of the riser, while the particle volume concentration decreases along the height of the riser. This stage is called turbulent fluidized bed, or circulating fluidized bed if the particles are returned to the riser (Figure 2-9e). Usually cyclone separators are applied for the separation of gas and solid phase at the outlet of a fluidized bed. If the fluidization velocity is further increased, all particle starts to move immediately according to the gas flow, which is called pneumatic transport (Figure 2-9e).

The abovementioned fluidization regimes are marked in Figure 2-10 as a function of the dimensionless particle diameter, d_p^* (Eq. 2-7) and of the dimensionless superficial gas velocity, u_0^* (Eq. 2-8). Furthermore, the different classes of particles are indicated according to the classification of Geldart [84].

$$d_p^* = d_p \left[\frac{\rho_f (\rho_p - \rho_f) g}{\mu^2} \right]^{\frac{1}{3}} \qquad (2\text{-}7)$$

$$u_0^* = u_0 \left[\frac{\rho_f^2}{\mu(\rho_p - \rho_f) g} \right]^{\frac{1}{3}} \qquad (2\text{-}8)$$

A hydrodynamic characterization of the operational range of the experimental investigation for carbonator and calciner is given in Chapter 3.5.

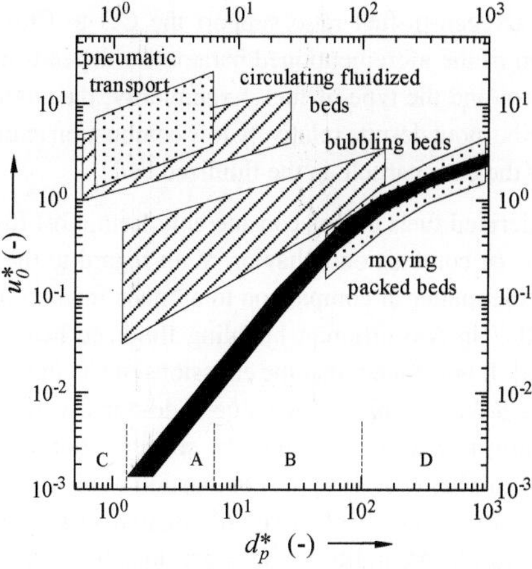

Figure 2-10: Fluidization states for gas/solid flow according to Grace et al. [86], extended from [83].

2.4 Fluidized Bed Combustion of Solid Recovered Fuel

The application of FB systems for the combustion of waste-derived fuels, such as SRF, is a well-established method. The moderate combustion temperatures in FB boilers reduce the formation of NO_x while the emissions of gaseous sulfur and chlorine species can be effectively limited by the feed of additives directly into the bed. In contrast to conventional solid fuels such as lignite or hard coal, SRF contains increased quantities of critical fuel constitutes such as chlorine, heavy metals and alkaline metals. Furthermore, the composition and quality of the fuel is highly heterogeneous, making it challenging to characterize the combustion behavior and the formation of gaseous emissions based on the pure elementary fuel composition. Experimental investigations are thus inevitable in order to assess the combustion of SRF in novel applications. With regard to the CaL process, the combustion behavior of SRF in the calciner is of concern, as it influences the sorbent deactivation and the upgrading requirement of the calciner CO_2-rich flue gas in the subsequent GPU. This chapter introduces the fundamentals related to the formation and mitigation of the major gaseous emission pollutants such as carbon monoxide (CO), nitrogen oxides (NO_x), sulfur dioxide (SO_2) and hydrogen chloride (HCl). Until now, no comprehensive experimental study on SRF combustion under various combustion atmospheres in a CFB furnace has been published.

2.4.1 Emission of Carbon Monoxide

During the combustion of solid fuels, CO is formed in the course of pyrolysis and char oxidation. The CO is then fully converted to CO_2 if process conditions allow the complete oxidation. Thus, the emission of CO can be attributed to the incomplete combustion of carbon based fuel species. In addition to environmental, health and safety concerns, the emission of CO in fact lowers to the combustion efficiency of the furnace. The mechanisms related to CO formation and oxidation have been largely studied. It is known, that a reasonably high combustion temperature and residence time along with an appropriate oxygen-to-fuel ratio support the CO to CO_2 conversion [87, 88]. The quantitative characterization of the aforementioned parameters depends on the boundary conditions set by the combustion system and the type of fuel. Exemplarily, the environment of a CFB furnace allows for a reasonable char burnout despite relatively low combustion temperatures as a consequence of a large residence time of the char particle in the fluidized bed.

The composition of waste-derived fuels differs considerably from most fossil fuels, which have been extensively studied in terms of combustion behavior. With regard to the formation mechanisms of CO, the large share of volatile matter in comparison to a minor fraction of fixed carbon plays a key role. The combustion of RDF in two different bubbling fluidized bed combustors was studied by Hernandez-Atonal et al. [89]. It was found, that the emissions of CO increase exponentially when the oxygen-to-fuel ratio approach the stoichiometric value. Additionally, the author concluded, that the use of secondary air injection is crucial to reduce the concentration of CO in the flue gas of the combustor. In another study of Sher et al. [90] the combustion of biomass, which is comparable with waste-derived fuels in terms of a high share of volatile matter, was studied in a 20 kW_{th} bubbling fluidized bed reactor. These results are in line with the previous findings that the emission of CO can be suppressed by an appropriate feed of secondary air. This was justified by the fact that there is a higher temperature in the freeboard, which is the main combustion zone for fuels with a high share of volatile matter. Contrary to the emission of CO, both studies indicate that a higher oxygen-to-fuel ratio favors the formation of NO_x.

2.4.2 Emission of Nitrogen Oxides

The formation of nitrogen oxides (NO_x) during the process of combustion of solid fuels has been widely studied. The main share of NO_x refers to NO, while the formation of NO_2 is of minor importance [91]. As another combustion product, N_2O is partly formed which is of concern as its greenhouse gas potential is approximately 200 times as high as that of CO_2 [3].

There are three routes responsible for NO_x formation (e.g. thermal, fuel, prompt), whereas the thermal route is of less importance in CFB combustion systems because of the moderated combustion temperature. It was found that the fuel-route represents the dominating NO_x formation pathway under combustion conditions in a CFB [88]. Here, the fuel-N that is released during the devolatilization is converted via the intermediate products NH_3 and HCN to NO_x or N_2. The share of fuel-N bound in the fixed carbon is oxidized to NO. The formation patterns of NO are, among others, strongly related to the form in which the nitrogen is bound in the fuel. For the case of SRF combustion, almost all fuel-N is released via devolatilization in the form of the intermediate products NH_3 and HCN.

While comparing oxyfuel and air combustion systems, the volumetric concentration of NO_x in the flue gas tends to be higher in the oxyfuel case, whereas the mass unit of NO_x per thermal duty of the boiler is lower. This is due to the absence of air-nitrogen in the oxidation agent and due to the fact of flue gas recirculation [92]. The latter effect causes the reduction of NO towards N_2 and N_2O in the lower region of the riser. Here, the reduction pathway to N_2O was found to be less than 5 % of all recirculated NO [93].

2.4.3 Emission of Sulfur Dioxide

The detailed oxidation scheme of sulfur is related the actual form in which the sulfur is present in the fuel. The chemical form of the fuel sulfur further determines whether it is released during devolatilization or during the combustion of fixed carbon. During the course of oxidation, either sulfur dioxide (SO_2) or sulfur trioxide (SO_3) is formed. Even though the formation of SO_3 is critical in terms of high and low-temperature corrosion processes, the main share of fuel sulfur is converted to SO_2 [94].

SO_2 can be captured by the sulfation of calcium compounds in the fluidized bed. Typically, limestone is utilized for in-bed desulfurization. In addition to the limestone that is actively fed to the bed, the presence of Ca-species in the fuel ash already lead to partial desulfurization. The effectiveness of this measure is related to the physical and chemical form of the Ca-species in the fuel ash. Eq. 2-9 shows the direct desulfurization pathway:

$$CaCO_{3(s)} + SO_{2(g)} + \frac{1}{2}O_{2(g)} \rightarrow CaSO_{4(s)} + CO_{2(g)} \tag{2-9}$$

Once the limestone absorbs the SO_2 in the calcined form, the desulfurization mechanism is referred to as indirect desulfurization. A highly porous particle structure is formed during the course of calcination, which favors the absorption of SO_2:

$$CaO_{(s)} + SO_{s(g)} + \frac{1}{2}O_{2(g)} \rightarrow CaSO_{4(s)} \tag{2-10}$$

As oxygen is required in the sulfation reaction, oxidizing conditions are a prerequisite for in-situ desulfurization [95]. The quantity of Ca-species relatively to the sulfur in the fuel is typically expressed as the molar Ca/S ratio. The chemical equilibrium in the $CaCO_3$-CaO system (see Chapter 2.5.2) determines whether the direct or indirect desulfurization mechanism is dominating.

Under combustion conditions without flue gas recirculation, the indirect desulfurization mechanism is prevalent, as the CO_2 partial pressure is relatively low (p_{CO2}: < 0.15). However, under conditions of an oxyfuel combustion system, there will be a mixture of both desulfurization mechanisms because of the higher CO_2 partial pressure (p_{CO2}: ~ 0.4 - 0.75). While in the oxyfuel combustion environment, the solid phase properties are less favorable for SO_2 absorption, the absence of air leads to higher concentrations of the gaseous combustion products (i.e. SO_2), which ultimately favors the SO_2 absorption.

In the work of Gomez et al. [96], the sulphur capture was investigated at a 30 MW_{th} CFB boiler in air and oxyfuel combustion atmospheres using an anthracite coal. It was found that the sulphur capture was higher in oxyfuel combustion mode. According to the authors, this is related to the higher residence time of SO_2 due to flue gas recirculation, thus sulfation is increased. The ideal temperature for desulphurization is given at a molar Ca/S ratio of 2.6 at approximately 880 - 890 °C under oxyfuel conditions. Li et al [97] investigated the combustion of Datong coal in a 0.1 MW_{th} CFB test rig at various combustion atmospheres. It was found that for a Ca/S ratio of 2.2, the sulfur capture was 70.2 % in oxyfuel combustion mode and 93.7 % in air combustion mode.

2.4.4 Emission of Hydrogen Chloride

Hydrogen chloride (HCl) represents the primary combustion product of chlorine species present in the fuel. Several studies indicate the ability of Cl-retention by Ca-species during the combustion of waste-derived fuels in FB combustion systems [98, 99]. Under the typical conditions of the oxyfuel CaL calciner, the reaction product tends to form molten phases, which significantly alters the reaction mechanisms [100]. The HCl absorption phenomenon by Ca-species in a FB is highly complex and multi-layered. From the literature, it is known that the dechlorination mainly depends on the temperature, gas atmosphere and the molar Ca/Cl ratio [101]. Under the typical conditions of FB boilers, the reaction mechanisms can be described by Eqs. 2-11 and 2-12 [102, 103]:

$$CaCO_{3(s)} + 2HCl_{(g)} \leftrightarrow CaCl_{2(s,l)} + H_2O_{(g)} + CO_{2(g)} \tag{2-11}$$

$$CaO_{(s)} + 2HCl_{(g)} \leftrightarrow CaCl_{2(s,l)} + H_2O_{(g)} \tag{2-12}$$

The emission behavior of HCl during the combustion of source-classified combustible solid waste in a fluidized bed combustor was investigated by Li et al. [104]. It was found that the emission of HCl can be effectively lowered by the addition of CaO to the furnace. The quantity of HCl in the flue gas was as low as 2.5 mg_{HCl}/m^3 under ideal dechlorination conditions. In the work of Ferrer et al. [105], the co-combustion of RDF with coal was investigated in a 100 kW_{th} and a 4 MW_{th} CFB furnace. The authors concluded that the Ca-species based absorption of SO_2 dominates over the absorption of HCl. This finding is in accordance with another study by Partanen et al. [106]. Until now, the HCl absorption behavior under typical conditions of a CaL calciner has not been investigated.

2.5 Carbonate Looping Process

The CaL process is a second generation post-combustion CO_2 capture process making use of the reversible carbonation-calcination reaction between lime and CO_2. The circulating sorbent stream is exposed to cyclic carbonation-calcination reaction regimes, whereby CO_2 is separated from a flue gas stream. The CaL process was initially proposed for the purpose of CO_2 capture in the framework of coal-fired power plants by Shimizu et al. in 1999 [107].

2.5.1   Fundamentals and Process Layout

Figure 2-11 provides a schematic of the CaL process principle. The CaL process uses two interconnected fluidized bed (FB) reactors, typically named carbonator and calciner. To allow for small plant dimensions, both reactors are usually designed as circulating fluidized bed (CFB) reactors. The flue gas of the upstream industrial process is introduced to the carbonator, where CO_2 is absorbed by solid CaO according to the exothermic carbonation reaction:

$$CO_{2(g)} + CaO_{(s)} \leftrightarrow CaCO_{3(s)} \tag{2-13}$$

Hereby, calcium carbonate ($CaCO_3$) is formed and heat is released ($\Delta H^0 = +/- 178.2$ kJ/mol). The partly carbonated sorbent is then transferred to the calciner. In the calciner, the temperature of the sorbent is raised in order to establish the conditions for calcination. Subsequently, $CaCO_3$ decomposes to CaO and CO_2 via the endothermic calcination reaction (see Eq. 2-13).

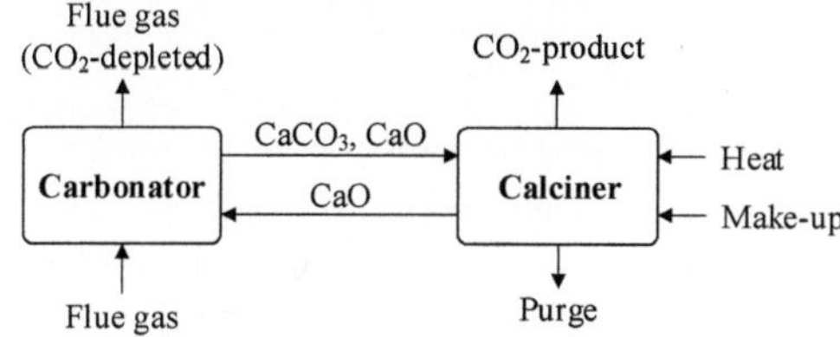

Figure 2-11: Schematic of the CaL process.

In order to account for the decreasing ability of the sorbent to absorb CO_2 and the dilution with ash, a constant flux of fresh limestone make-up is fed to the process, while spent sorbent and ash are discharged.

The operation temperature of the carbonator is determined by the reaction kinetics of the carbonation reaction and the chemical equilibrium. Approximately 650 °C are considered an ideal value for reasonable fast reaction kinetics and theoretically maximum CO_2 absorption rates close to the chemical equilibrium [108]. The operation temperature of the calciner of approximately 900 °C is required to ensure a complete calcination of the circulating sorbent even under high CO_2 partial pressure [109, 110].

The carbonator represents an exothermic system, as heat is released in the course of the carbonation reaction. Additionally, heat is introduced by the hot solid stream coming from the calciner. In contrast to that, the calciner represents an endothermic system, due to the heat consumption of the calcination reaction and the heat required for temperature increase of the incoming material streams.

The ability of the circulating sorbent to absorb CO_2 in the carbonator decreases with increasing numbers of carbonation-calcination cycles. This multi-layered phenomenon is mainly associated with the sintering of the particle pores and the deactivation of the sorbent due to the formation and enrichment of inert species (see Chapter 2.5.4).

The layout of the CaL process is divided according to the method for the supply of heat to the calciner. It is common to distinguish between concepts for a direct and for an indirect heat supply to the calciner. Figure 2-12 shows the schematics of the directly heated CaL process (a) and of the indirectly heated CaL process (b).

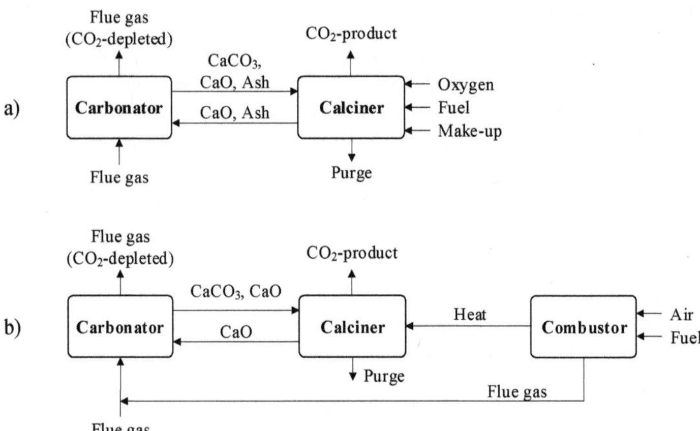

Figure 2-12: Schematic of the directly heated CaL process (a) and of the indirectly heated CaL process (b).

The directly heated CaL process design represents the most developed configuration. It is based on the combustion of supplementary fuel directly in the calciner. In order to achieve a highly-concentrated CO_2 stream at the outlet of the calciner, the fuel is combusted under oxyfuel conditions. Oxygen from the ambient air is thus replaced by technically pure oxygen, typically supplied by an on-site cryogenic ASU. As usual for oxyfuel combustion systems, flue gas needs to be recirculated for the dilution of the oxygen in the oxidation agent [49]. The directly heated CaL process has been investigated intensively within the past decades and is currently ranked at TRL6 [53]. The process has been demonstrated in relevant environment and scale. Its development is further sustained by the similarities between the CaL reactors and the already state-of-the-art CFB combustion systems.

Due to the direct contact of circulating sorbent and combustion residues of the supplementary fuel in the calciner, sorbent deactivation is facilitated in the directly heated CaL process design. This is related to the enrichment of inert species in the solid phase and to the exposure of the Ca-species to the sulfur present in the supplementary fuel. Furthermore, the in-bed combustion induces local temperature peaks that further contribute to particle sintering in the calciner. In addition to the sorbent-related disadvantages, the operation of the ASU represents the main drawback as it significantly increases the auxiliary power consumption of the directly heated CaL process.

In order to avoid some of the aforementioned weaknesses of the directly heated CaL process, novel process designs have been developed. Figure 2-12b shows a schematic of the indirectly heated CaL process whereby the heat required in the calciner is supplied indirectly from an external combustion chamber, where the supplementary fuel is combustion in a conventional air environment [111]. The operation of an ASU is thus avoided, which increases the net electrical efficiency noteworthy. Furthermore, there is no direct contact of supplementary fuel species and circulating sorbent, which hinders the sorbent deactivation. In order to allow the heat transfer to the calciner, the external combustion chamber needs to be operated at approximately 1,000 °C. The resulting flue gas is fed to the carbonator for the CO_2-depletion in addition to the flue gas of the upstream industrial unit.

Several principles have been proposed for the heat transfer from the external combustor to the calciner, such as the heat transfer through metallic walls [111], the utilization of a heat carrier material e.g. Al_2O_3, Fe_3O_4 or $CaO/CaCO_3$ [112,113] or the application of heat pipes as a heat transfer mean

[114]. All of the three concepts are prone to specific advantages, disadvantages and challenges. The only concept that has been successfully tested on TRL5 at this point is the indirect heat transfer by means of heat pipes [115].

2.5.2   Chemical Equilibrium in the CaCO₃-CaO System

The characteristics of the CaL process are governed by the chemical equilibrium conditions in the CaCO₃-CaO system. There are several empirical correlations published for the calculation of the chemical equilibrium CO_2 concentration dependent on the prevailing temperature [116-119]. In Eq. 2-14, an exemplary correlation from Garcia-Labiano et al. [116] is given:

$$y_{CO2,eq} = 4.137 \cdot 10^7 e^{-\frac{20474\,K}{T}}$$

(2-14)

Here, $y_{CO2,eq}$ is the CO_2 concentration at chemical equilibrium conditions at a given temperature, T. Figure 2-13 shows the resulting curve at chemical equilibrium that divides the carbonation and calcination reaction regimes. The figure includes the significant temperature field for the CaL process between 600 and 900 °C.

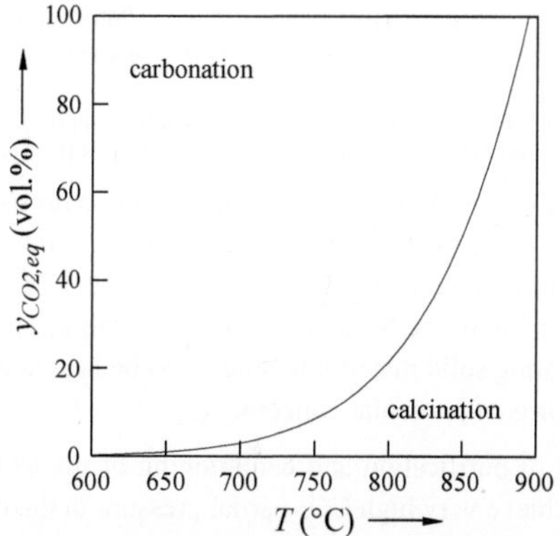

Figure 2-13: Chemical equilibrium of the carbonation-calcination system at atmospheric pressure.

The theoretical maximum CO_2 absorption efficiency in the carbonator, $E_{carb,eq}$, is relevant for the CaL process:

$$E_{carb,eq} = \frac{F_{CO2,abs,eq}}{F_{CO2,in}}$$

(2-15)

In the above equation, $F_{CO2,abs,eq}$ is the molar flux of CO_2 being absorbed by the solid phase at chemical equilibrium and, $F_{CO2,in}$ is the molar flux of CO_2 contained in the flue gas at the carbonator inlet.

The inlet conditions are given by the corresponding host plant (i.e. CO_2 concentration, volumetric gas flow), whereas the molar flux at the outlet is dependent on the chemical equilibrium. The theoretical maximum CO_2 absorption efficiency within the carbonator increases accordingly with lower carbonator temperatures or higher CO_2 inlet concentrations. Contrary to that, the kinetics of the carbonation reaction favor higher operation temperatures. It was shown that a carbonator temperature below 550 °C is insufficient for typical CaL applications [120]. On the other hand, a carbonator temperature of approximately 650 °C has been identified as the ideal compromise between carbonation reaction kinetics and chemical equilibrium limitations [121].

Figure 2-14 shows the theoretical maximum CO_2 absorption efficiency in the carbonator for different CaL process back-end applications (e.g. cement plant, coal-fired power plant, WtE plant).

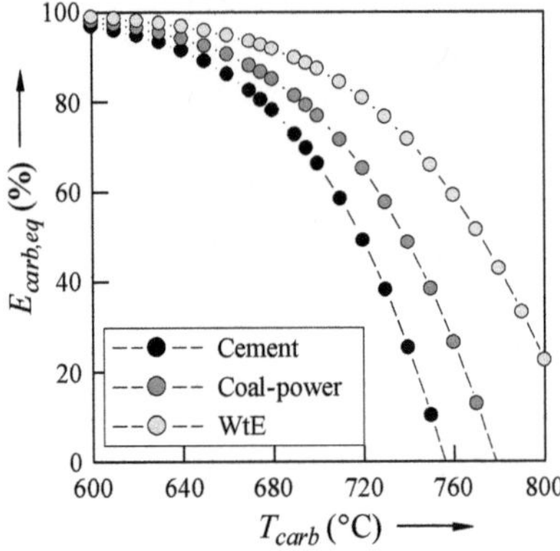

Figure 2-14: Theoretical maximum CO_2 absorption efficiency at chemical equilibrium for different CaL applications (Cement: $y_{CO2,in}$ = 30 vol.%, Coal-power: $y_{CO2,in}$ = 14 vol.%, WtE: $y_{CO2,in}$ = 9.5 vol.%).

The theoretical maximum CO_2 absorption efficiency in the carbonator is significantly influenced by the given CaL application. In order to allow for a CO_2 absorption efficiency of approximately 80 %, the carbonator needs to be operated at approximately 675, 688 and 722 °C for the applications of a WtE plant, coal-fired power plant or the back-end integration into a cement plant, respectively. Due to the high amount of circulating solid material that needs to be heated in the calciner, the operation temperature of the carbonator is of particular concern.

In order to keep the efforts of purification and conditioning of the CO_2 product stream as low as possible, it is desirable to achieve very high CO_2 partial pressure in the calciner. Thus, the operation temperature of the calciner needs to be in the range of 850 to 900 °C to ensure the calcination of the circulating sorbent. It is important to note that the temperature of the calciner is also crucial for sorbent deactivation, as higher calcination temperatures facilitates the sintering phenomenon of the solid particles. Additionally, higher calcination temperatures cause a higher fuel and oxygen consumption. The latter aspect significantly increases the auxiliary power consumption of the CaL process.

2.5.3 Carbonation Reaction

The carbonation reaction is one of the key phenomena of the CaL process. It determines the CO_2 absorption efficiency in the carbonator and the total CO_2 capture rate of the CaL process. In the course of this section, the carbonation reaction regimes and the effect of water vapor on the carbonation reaction are explained.

2.5.3.1 Carbonation Reaction Regimes

The carbonation reaction proceeds in two distinguished reaction regimes. In the beginning of the reaction, a fast CO_2 uptake proceeds until a critical $CaCO_3$ product layer has been formed at the inner surface of the CaO particle. The critical product layer thickness has been experimentally determined to approximately 50 nm [122]. Thereafter, the further proceeding of the carbonation reaction is limited by the diffusion of CO_2 through the product layer. Whereas the first mentioned fast carbonation regime is limited by the carbonation reaction kinetics, the subsequent slow carbonation regime is limited by the diffusion resistance of the product layer towards CO_2. It was found that this reaction phenomenon occurs independently of the number of carbonation-calcination cycles [123]. The molar conversion of a CaO particle, X is calculated as follows:

$$X = \frac{n_{CaCO3}}{n_{Ca}} \tag{2-16}$$

Here, n_{CaCO3} is the molar quantity of $CaCO_3$ and, n_{Ca} is the total molar quantity of Ca in a given sample, respectively. Figure 2-15 shows an exemplary plot of the carbonation sequence of two samples carried out in a thermogravimetric analyzer (TGA). The associated experimental setup is described thoroughly in Chapter 3.4.1.4. A sample of the make-up limestone (Fresh) and of a highly cycled sorbent sample extracted from the 1 MW_{th} pilot plant (Used) was investigated. It needs to be noted that the quantitative course of the curve progression depends on the type of limestone and on the prevailing process conditions, while the qualitative course is similar among all types of limestone that were considered for the application in the CaL process until now [124].

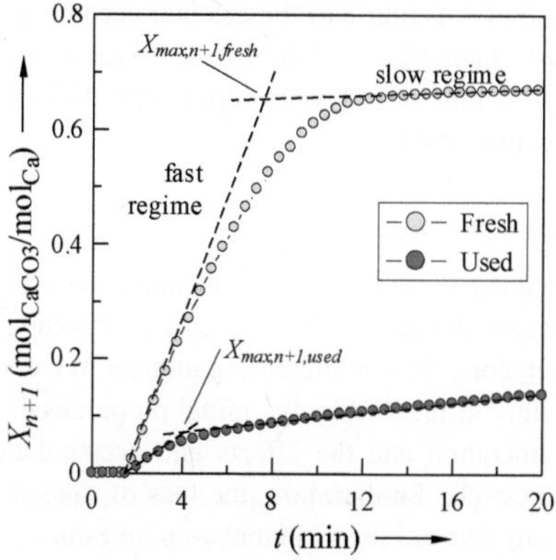

Figure 2-15: TGA carbonation sequence of a fresh and a spent limestone sample.

The turning point of the fast and of the slow reaction regime is typically referred to as maximum CO_2 carrying capacity, X_{max}. It is graphically derived from the intersection of two parallel lines to the carbonization curve of each reaction regime. Even though there is a further CO_2 absorption in the slow reaction stage, the fast reaction regime is considered as relevant for technical applications of the CaL process. In case of Figure 2-15, the maximum CO_2 carrying capacity is approximately 0.650 mol$_{CaCO3}$/mol$_{Ca}$ and 0.085 mol$_{CaCO3}$/mol$_{Ca}$ for the fresh and for the used sorbent sample. The turning point between the two reaction regimes is more difficult to determine in case of the used sorbent sample as the sintering of the particle surface significantly reduces the overall carbonation rate in the fast regime. Furthermore, the used sorbent sample consists of a mixture of particles with a varying number of carbonation-calcination cycles.

2.5.3.2 Effect of Water Vapor

In the standard CaL process layout, the circulating sorbent is exposed to water vapor in the carbonator and in the calciner. The presence of water vapor in the carbonator is related to the composition of the flue gas of the upstream process. For the calciner, the presence of water vapor is associated with the combustion of supplementary fuel and the recirculation of wet flue gas for oxygen dilution. Depending on whether the sorbent is exposed to water vapor during the carbonation or calcination reaction step, the effects differ.

It is known, that the presence of water vapor in the gas phase during the carbonation has a positive effect on the CO_2 absorption. The explanation of the positive water vapor effect has been controversially discussed in literature and is attributed to two mechanisms. The first mechanism is relevant for the fast reaction stage and implies the catalytic effect of H_2O on the carbonation reaction by the formation of $Ca(OH)_2$ [125]. The second mechanism enhances the carbonation during the diffusion controlled reaction regime by breaking the carbonate product layer [126]. It was found that the influence of water vapor during the slow reaction dominates. Accordingly, the turning point between the two reaction regimes has been extended towards higher molar conversion degrees, which enhances the CO_2 uptake capacity during fast carbonation reaction stage [127].

The presence of water vapor during the calcination step favors the sintering of the particles [128, 129]. The activity decreases due to the effect of surface reduction. On the other hand, it is important to note that due to the presence of water vapor, the partial pressure of CO_2 is lowered, which further allows a reduced pf the operation temperature of the calciner according to the chemical equilibrium of the CaO-$CaCO_3$ system (see Chapter 2.5.2). Ultimately, it was experimentally proven that under the aforementioned conditions, i.e. presence of water vapor and reduced calcination temperatures, the sorbent is less prone to deactivation [130].

2.5.4   Sorbent Deactivation Mechanisms

The deactivation of the circulating sorbent is a key phenomenon within the CaL process which significantly influences the process characteristics. The term deactivation emphasizes the loss of CO_2 carrying capacity of the sorbent along with an increasing number of carbonation-calcination reaction cycles. The deactivation is mainly influenced by the initial properties of the limestone, the sintering of the particle pores during operation and the effects and accumulation of fuel-species that are introduced to the solid looping cycle. Furthermore, the loss of fines as a consequence of attrition phenomena leads to a continuous demand for fresh make-up limestone.

2.5.4.1   Sintering of the Particle Pores

The deactivation of the sorbent, i.e. the reduction of sorbent conversion at the turning point from the fast to the slow carbonation reaction regime proceeds with an increasing number of carbonation-calcination cycles. More precisely, the numbers of micro pores decreases, while more macro pores are being formed. As a consequence, the specific surface area of the sorbent that is available for CO_2 absorption decreases [129, 131, 132]. From the literature, it is known that the phenomenon of sintering occurs mainly during the high-temperature sorbent regeneration in the calciner [133]. Consequently, operation conditions of the calciner are of concern for the progress of sintering. The gas atmosphere (i.e. presence of H_2O and CO_2), the operation temperature and the residence time of particle in the calciner are of particular interest. Blamey et al. [134] found that a decline in sorbent activity is forced at higher calcination temperatures (1,000 °C) in comparison to moderate calcination temperatures (840 °C). An ideal compromise between sorbent deactivation and calcination efficiency needs to be found since a temperature sufficiently far away from chemical equilibrium in the CaO-$CaCO_3$ system is required in the calciner. Furthermore, it is known that the presence of CO_2 in the gas phase during calcination increase the rate of sintering [135].

The deactivation phenomenon has been widely discussed within the past years by several research groups worldwide. Most of the studies rely on experimental investigations carried out in TGA devices that mimic the cyclic carbonation-calcination reaction under various process conditions. It was found that the extent of deactivation per carbonation-calcination cycle is mainly a function of the calcination temperature and gas atmosphere, as well as of the residence time of the particles in the calciner. Equation 2-17 gives a widely used empirical correlation for the deactivation of the limestone up to 500 carbonation-calcination cycles [123].

$$X_{max,N} = \frac{1}{\frac{1}{1 - X_r} + kN} + X_r$$

(2-17)

Here, $X_{max,N}$ states the maximum molar carbonation conversion at, N complete carbonation-calcination cycles, X_r states the residual conversion capacity of the sorbent at an infinitely number of cycles, and k states the deactivation constant. The residual conversion capacity and the deactivation constant can be seen as fitting parameters in order to account for different types of limestone. For natural limestone and typical calcination conditions of the CaL process ($T_{calc} < 950\ °C$, $t_{calc} < 20$ min), a residual conversion capacity of 0.075 and a deactivation constant of 0.52 were identified as best fit [123].

The molar conversion of the sorbent based on Eq. 2-17 is shown in Figure 2-16 for the best fit parameter (Base) and for best and worst deactivation conditions (Best, Worst) as identified by TGA testing [123].

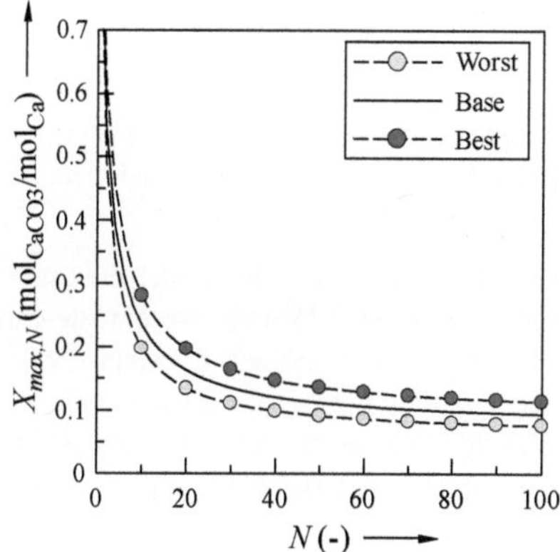

Figure 2-16: Sorbent activity as a function of the cycle number for different fitting parameters of Eq. 2-17 (Worst: $X_r = 0.06$, $k = 0.624$; Base: $X_r = 0.075$, $k = 0.52$; Best: $X_r = 0.09$, $k = 0.416$).

The deactivation phenomenon is strongly pronounced within the first 20 cycles. Thereafter, the further sorbent deactivation is limited and tends towards the residual sorbent conversion capacity. For large-scale applications of the CaL process, it is reasonable to keep the sorbent activity close to its residual conversion capacity, as this limits the need for additional make-up limestone requirement.

2.5.4.2 Sulfur Induced Sorbent deactivation

The activity of the sorbent is also reduced by sulfur species induced deactivation mechanisms. Sulphur is introduced to the CaL process via two pathways. First, in the form of sulfur oxides (mainly

SO_2) contained in the flue gas of the upstream process and second, as fuel sulfur contained in the supplementary fuel that is fired in the calciner.

The concentration of SO_2 in the flue gas of the host process is typically controlled either by primary (e.g. in-bed desulfurization) or secondary measures (e.g. flue gas cleaning unit) of the upstream plant. The concentration is thereby lowered according to local stack emission limits. The quantity of fuel sulfur depends on the type of supplementary fuel. Given the characteristics of the oxyfuel calciner, the fuel sulfur will be oxidized to SO_2 or SO_3 during the process of combustion. Ca-species react with SO_2 either directly (Eq. 2-9), or indirectly (Eq. 2-10) depending on the prevailing CO_2-partial pressure. As a product of both reactions, calcium sulfate ($CaSO_4$) is formed (see Chapter 2.4.3).

In conclusion, the quantity of flue gas sulfur is avoidable by an appropriate setting of the flue gas cleaning line. In contrast, the share of fuel sulfur and the related deactivation cannot be avoided, rather only be limited by an appropriate fuel choice. The decomposition of $CaSO_4$ occurs at a temperature above 1,100 °C [132]. $CaSO_4$ thus remains stable in the temperature range that is typically considered for the CaL process (T_{CaL}: 600 - 950 °C). The share of $CaSO_4$ represents an inert irreversible consumed fraction of Ca-species that is no longer available for CO_2 absorption. Furthermore, it increases the thermal requirement of the calciner as it participates in the cyclic temperature increase and decrease cycles of the sorbent.

Another deactivation mechanism of the $CaSO_4$ formation is the blockage of the outer particle pores as the molar volume of $CaSO_4$ ($V_{m,CaSO4}$: 45.7 cm³/mol) exceeds that of CaO ($V_{m,CaO}$: 15.0 cm³/mol) [136]. This mechanism causes a continuous plugging of the pores which reduces the available inner surface area for CO_2 absorption.

2.5.4.3 Effect of Chlorine

Chlorine might be introduced to the CaL process via two pathways. First as component of the flue gas of the host process to be decarbonized in the carbonator and second as part of the supplementary fuel that is burnt in the calciner.

The first part is insignificant in almost all conventional fuel-fired processes as the chlorine mass fraction of fossil fuels is typically well below 0.25 wt.%. Once waste-derived fuels are considered of being co- or mono-fired in the host process or in the CaL calciner, chlorine is largely converted to hydrogen chloride (HCl). In case of the host process, the concentration of HCl in the flue gas is limited either by primary (e.g. in-bed dechlorination) or secondary measures (e.g. flue gas cleaning unit), while in the calciner the sorbent is directly exposed to the gaseous HCl. Thus, the latter share cannot be avoided.

The HCl absorption phenomenon by Ca-species in a CFB is highly complex and multi-layered. Under typical conditions of FB boilers, the reactions mechanism can be described by Eqs. 2-11 and 2-12 (Chapter 2.4.4).

The effect of HCl on the carbonation-calcination reaction cycle has been rarely discussed until now. The influence of HCl absorption on the sorbent activity within the CaL process was assessed by means of TGA in a study from Symonds et al. [137]. It was found that the presence of HCl during the carbonation step might lead to a more active sorbent due to the formation of larger pores during carbonation. In another study, a fixed bed reactor system was applied to expose Ca-species to HCl during the cyclic CO_2 absorption. Here, the discontinuous addition of HCl during the carbonation step was also identified as a method of improving the CO_2 capture capacity [138]. In conclusion, it can be staid that the presence of HCl during the carbonation step improves sorbent activity, while on the other hand, part of the formed $CaCl_2$ might decompose under conditions of a CaL calciner. Further investigations are required at this point to assess the fate of chlorine in the CaL process.

2.5.4.4 Enrichment of Impurities in the Solid Phase

Solid impurities (e.g. SiO_2, Fe_2O_3, Al_2O_3, $MgCO_3$) are introduced to the solid looping system by the mineral matter of the supplementary fuel fired in the calciner and by the limestone make-up feed to the process. The enrichment of impurities in the circulating solid phase reduces the activity because of two mechanisms.

Most impurities remain inert at the process conditions of the CaL process. As they participate in the solid material circulation between carbonator and calciner, the heat demand of the calciner and the cooling duty of the carbonator increase simultaneously. The dimensions of the CaL unit thus increase, while the specific CO_2 carrying capacity of the circulating sorbent decreases.

The other deactivation mechanism is related to the formation of other calcium compounds (e.g. $Ca_2 \cdot SiO_4$, $CaO \cdot Al_2O_3$, $2CaO \cdot Al_2O_3 \cdot SiO_2$) [139]. These compounds bind Ca-species which cannot contribute in the process of CO_2 absorption. In addition, the sintering of the particle surface is forced, as these compounds typically show a higher molar mass compared to $CaCO_3$.

2.5.4.5 Particle Abrasion and Attrition

Particle attrition is the result of mechanical, thermal and chemical stresses that arise in and between the particles during the movement in a fluidized bed. It is defined as the degradation of bed material and leads to a change in size and number of particles [140]. The attrition phenomenon depends on a number of parameters related to the design of the system, the properties of the solid phase, and the surrounding environment [141].

Knowledge of the sorbent abrasion characteristics within a fluidized bed process is important. Due to abrasion effects, the particle size reduces, while smaller particles are being formed. As a consequence, more fines are entrained throughout the cyclone separators of the CFB. In order to keep the CFB solid inventory constant, fresh sorbent needs to be fed to process. Among others, this phenomena causes several technical disadvantages due to the higher solids loading of the flue gas. These are for instance a higher pressure drop or the increasing erosion of heat exchanger surfaces. Additionally, economics decline due to the costs related to the procurement and disposal of additives and bed material. Another aspect is associated with the change of specific surface area in the solid phase, which particularly influences process performance of chemical processes.

There is already substantial knowledge of limestone attrition behavior due to its application as desulfurization agent in fluidized bed furnaces [136, 142]. Nevertheless, the CaL process conditions require an extension of the current state of research to allow the prediction of sorbent attrition under relevant process conditions.

It is known that the presence of a carbonate, or of a sulfate product layer increases the mechanical strength of Ca-particle [143, 144]. Among these two product layer options, the influence of the sulfate is of less importance, as its appearance is usually less than one order of magnitude compared to the carbonate layer under conditions of the CaL process. The carbonate layer is ideally fully formed on the Ca-particle at the outlet of the carbonator, while almost only oxide is present at the outlet of the calciner.

A major cause of particle fragmentation during the process of calcination is related to the release of gaseous CO_2 within the porous particle structure and to the rapid temperature increase while the sorbent is introduced to the calciner. It was already found that the first effect dominates [145]. Thus, one could conclude that a higher calcination temperature leads to an increased attrition rate because of faster CO_2 release. Furthermore, the first calcination of the limestone make-up is known to cause the highest attrition rates [146], as the whole particle initially consists of calcium carbonate. Consequently, the attrition related to the calcination step decreases with a higher carbonation/calcination cycle number since the ability of the particle to absorb CO_2 decreases simultaneously.

2.5.5 Pilot-Scale Investigation of the Carbonate Looping Process

Within the past years, the continuous CO_2 capture by means of the CaL process has been successfully demonstrated in various test rigs up to megawatt-scale worldwide. The development of the CaL process is boosted by similarities between the CaL CFB reactors and the already well understood fundamentals of CFB combustion systems. Experimental results were reported from several test facilities varying in process layout and equivalent thermal size. Table 2-4 presents an overview and key characteristics of worldwide CaL process test facilities. Relevant CaL pilot plants are described in more detail below.

Table 2-4: Overview and key characteristics of worldwide CaL process test facilities.

Place	Size (kW$_{th}$)	Flue gas source	Process configuration[2]	Reference
Stuttgart (GER)	10	Synthetic	Carb: CFB, BFB; Calc: CFB, BFB	[147, 148]
Tsinghua (CH)	10	Synthetic	Carb: BFB; Calc: BFB	[149]
Cranefield (UK)	25	Synthetic	Carb: EF; Calc: BFB	[150]
INCAR-CSIC (ES)	30	Synthetic	Carb: CFB: Calc: CFB	[151, 152]
CANMET (CAN)	75	Synthetic	Carb: MB, BFB; Calc: CFB	[120, 153]
Ohio (USA)	120	Synthetic	Carb: EF; Calc: Rotary kiln	[154]
Stuttgart (GER)	200	Synthetic	Carb: EF, BFB; Calc: CFB	[121]
INCAR-CSIC (ES)	300	Synthetic	Carb: CFB; Calc: CFB	[155, 156]
Darmstadt (GER)[3]	300	Synthetic	Carb: CFB; Calc: BFB Combustor: BFB	[157, 158]
Darmstadt (GER)	1,000	Comb. chamber	Carb: CFB; Calc: CFB	[159, 160]
Oviedo (ES)	1,700	Power plant	Carb: CFB; Calc: CFB	[161, 162]
Heping (TWN)	1,900	Cement plant	Carb: BFB; Calc: Rotary kiln	[163, 164]

<u>300 kW$_{th}$ indirectly heated pilot plant at Darmstadt, Germany</u>

The 300 kW$_{th}$ pilot plant at the Technische Universität Darmstadt in Germany is the only one of its kind with an indirectly heated calciner. The pilot plant consists of a CFB carbonator (d: 0.25 m, h: 8 m), a BFB calciner (l: 1.05 m, w: 0.3 m) and BFB combustion chamber (l: 1.05 m, w: 0.3 m). The combustion chamber is heated by firing propane in air atmosphere. The calciner and the combustion chamber are designed as one single component separated by a gas tight middle wall. The heat transfer from the combustion chamber to the calciner is implemented via 72 heat pipes using pure sodium as working fluid. The total heat transfer capacity accumulates to approximately

---

[2] CFB: Circulating fluidized bed, BFB: Bubbling bed, EF: Entrained flow, MB: Moving bed
[3] Indirectly heated calciner, heat transfer via sodium heat pipes from an external combustor

210 kW$_{th}$. The pilot plant has been operated for more than 340 hours, CO_2 absorption rates of up to 90 % from a synthetic flue gas stream were achieved [115, 157].

1.0 MW$_{th}$ pilot plant at Darmstadt, Germany

At the Technische Universität Darmstadt in Germany, a 1.0 MW$_{th}$ CFB test rig is operated. It consists of a combustion chamber for pulverized solid and gaseous fuels, as well as two CFB reactors (d_{R1}: 0.6 m, h_{R1}: 8.6 m; d_{R2}: 0.4 m, h_{R2}: 11 m). A unique feature of this unit is its diversity, as it is modifiable to conduct pilot-scale tests of various fluidized bed based processes. In this regard, processes such as CaL post-combustion CO_2 capture, chemical looping combustion (CLC) [165, 166] or FB gasification [167, 168] have been investigated extensively. Additionally, stand-alone CFB combustion tests have been carried out using hard coal, lignite, biomass or SRF as fuels [169].

The pilot plant was erected in 2010. Thereafter, the commissioning and first batch tests associated with the cyclic carbonation-calcination of limestone in a CFB reactor were carried out [159]. The next steps include the operation of the coupled CFB system aiming on continuous CO_2 capture with sorbent regeneration in the calciner, operated with oxygen-enriched air. Here, propane and coal were utilized as supplementary fuels [170, 171]. In the course of the research projects *LISA II* and *SCARLET*, the pilot plant was extended by a wet flue gas recirculation at the calciner [172]. In this new setting, the CaL pilot plant has been operated for more than 2,000 hours, including 1,500 hours under representative CO_2 capture conditions. This implies the decarbonization of a real flue gas in the carbonator and the operation of the calciner under realistic oxyfuel conditions with a wet flue gas recirculation. Over a wide range of process settings, it was found that CO_2 capture rates over 90 % are feasible, while using hard coal and lignite as supplementary fuels in the calciner. Part of the investigations dealt with novel coupling concepts and coupling components for the CFB reactors [160, 173, 174]. The experimental investigations as part of this thesis were also carried out at this unit (see Chapter 3.1).

1.7 MW$_{th}$ pilot plant at Oviedo, Spain

At Oviedo in north Spain there is a 1.7 MW$_{th}$ CaL test facility attached to the 50 MW$_e$ anthracite coal fired power plant La Pereda. The flue gas to be decarbonized in the carbonator is extracted right after the flue gas cleaning system of the host power plant. The pilot plant is operated by the Hunosa Company and scientifically supervised by the Institute of Carbon Science and Technology - Higher Council for Scientific Research (INCAR-CSIC). The reactor layout consist of a CFB carbonator (d: 0.65 m, h: 15 m) and a CFB calciner (d: 0.75 m, h: 15 m) as well as a recarbonator. Contrary to the aforementioned CaL pilot plant, the calciner of this unit uses a synthetic gas mixture for fluidization.

Major findings from this facility include the successful demonstration of an enhancement of the sorbent CO_2 carrying capacity due to an intermediate recarbonation step [175]. Doing so, the sorbent is exposed to a highly concentrated CO_2 stream ($y_{CO2,in}$ = 50 - 70 vol.%) in the carbonator loop seal for up to 3 minutes at approximately 750 - 800 °C. By this approach, the CO_2 carrying capacity was found to increase by up to 10 %-points. In another work, the calciner was operated at extreme oxyfuel conditions [176]. Here, the feasibility of calciner operation with an oxygen concentration of up to 75 vol.% in the oxidation agent was successfully proven. Due to the endothermic nature of sorbent calcination in the calciner, it is possible to enhance the oxygen in the oxidant in contrast to the conditions of a CFB oxyfuel boiler. This allows for a lower calciner dimension and energy savings. In another study, the dynamic CaL process behavior was investigated [177]. The authors demonstrated that part load operation conditions in terms of carbonator flue gas load of down to 50 % are feasible without the need for pilot plant modifications. The dynamic evolution of the CO_2 carrying capacity of the sorbent was assessed in another publication [161]. Similar to the previous study, this work aims on the application of the CaL process in the framework of flexible operated power plants. Based on experimental data, a particle population model was derived and successfully validated.

<u>1.9 MW$_{th}$ pilot plant at Heping, Taiwan</u>

The largest CaL pilot plant is located at Heping in Taiwan. It is operated under the responsibility of the Industrial Technology Research Institute (ITIR). The test facility is attached to a cement plant, which delivers the flue gases for the experimental investigations. The carbonator is designed as a BFB reactor (d: 3.3 m, h: 4.2 m), whereas the calcination of the sorbent is carried out in diesel-oxyfuel-fired rotary kiln calciner (d: 0.9 m, l: 5 m). Until now, the unit has been operated for more than 600 hours, whereof 300 hours were in continuous CaL operation under varying conditions [163, 164].

2.5.6 Carbonate Looping Process Modelling

The CaL process is still an emerging technology. The up-scaling and efficient design to an industrial size is thus faced with several challenges. Process modelling and simulation tools are a matter of importance to understand the potential for optimization and effective process control and design strategies. Within the CaL process, the deactivation of the circulating sorbent and the related CO_2 absorption efficiency in the carbonator are key parameters. This section provides an overview on the approaches that have been published regarding the modelling of the sorbent activity and CO_2 absorption in the fluidized bed carbonator.

2.5.6.1 Sorbent Activity

The characteristics of the sorbent activity in the CaL process have been widely studied (see Chapter 2.5.4). Most sorbent deactivation modelling approaches were derived from experimental TGA or fixed-bed reactor system analysis that mimic the cyclic carbonation-calcination of a sorbent sample. The deactivation patterns of the sorbent were investigated over a wide range of CaL process conditions and applied for the development empirical deactivation formula.

An early estimate from Abanades and Alvarez [178] is given Eq. 2-18. Here, the maximum CO_2 carrying capacity, $X_{max,N}$ is derived from the number of complete carbonation-calcination cycles, N and the limestone-specific fitting parameter f_m and f_w ($f_m = 0.77$ and $f_w = 0.17$ for natural limestone).

$$X_{max,N} = f_m^N(1 - f_w) + f_w \qquad (2\text{-}18)$$

In the second-order deactivation approach by Grasa and Abanades (see Eq. 2-19) [123], the maximum CO_2 carrying capacity calculation is based on a residual conversion capacity at an infinite number of complete carbonation/calcination cycles, X_r, and on the sorbent decay constant, k. Several TGA studies indicate that the deactivation mechanism is well reproduced by a deactivation constant (k) of 0.52 and a residual conversion capacity (X_r) of 0.075.

$$X_{max,N} = \cfrac{1}{\cfrac{1}{1 - X_r} + kN} + X_r \qquad (2\text{-}19)$$

Another semi-empirical correlation of Li et al. [179] is given in Eq. 2-20. The maximum CO_2 carrying capacity is determined by 5 fitting parameters (a_1, a_2, b, f_1, f_2) derived from TGA testing of three different sorbent types such as limestone, dolomite and synthetic mayenite-based sorbent.

$$X_{max,N} = a_1 f_1^{N+1} + a_2 f_2^{N+1} + b \tag{2-20}$$

In addition to the sintering phenomenon as the main cause for sorbent deactivation, the formation of $CaSO_4$ and the related blockage of particle pores is of concern. In order to incorporate this effect into the deactivation formulation, Romano [180] modified Eq. 2-19 based on the experimental data from Grasa et al. [181]. The deactivation constant and the residual conversion capacity of the sorbent were described as a function of the molar sulfation level of the sorbent, X_{CaSO4}:

$$k_{r,new} = k_r(1 + 0.2962 * X_{CaSO4}) \tag{2-21}$$

$$X_{r,new} = X_r(1 - 1.1536 * x_{CaSO4}), for\ X_{CaSO4} \leq 0.5 \tag{2-22}$$

$$X_{r,new} = X_r(-0.4230 - 3.076 * x_{CaSO4}), for\ X_{CaSO4} > 0.5 \tag{2-23}$$

Until then, all sorbent activity decay equations had been based on the assumption that the circulating sorbent shows the same number of complete carbonation-calcination cycles. However, in a real CaL application, the sorbent particles experience show a varying number of complete carbonation-calcination cycles, and additionally a different number of incomplete carbonation-calcination cycles. There will be a reaction age distribution for the sorbent. According to Abanades [182], the fraction of particles, r_N that have been circulated N times through the carbonation-calcination loop can be determined by Eq. 2-24:

$$r_N = \frac{F_0 F_R^{N-1}}{(F_0 + F_R)^N} \tag{2-24}$$

Based on the work of Rodriguez et al. [131], the average CO_2 carrying capacity of the sorbent in a continuously CaL process, X_{ave}, is calculated as follows:

$$X_{ave} = \sum_{N_{age}=1}^{N_{age}=\infty} r_{N,age} X_{N,age} \tag{2-25}$$

Where the reaction age of the sorbent, N_{age}, is calculated according to Eq. 2-26, which is based on Eq. 2-19. The fraction of particles in the corresponding reaction age, $r_{N,age}$, is derived by Eq. 2-27, while taking into account the degree of incomplete carbonation (f_{carb}) or calcination (f_{calc}), according to Eqs. 2-28 and 2-29.

$$N_{age} = \frac{1}{k}\left(\frac{1}{X_{N,age} + X_r} - \frac{1}{1 - X_r}\right) \tag{2-26}$$

$$r_{N,age} = \frac{(r_0 + \frac{F_0}{F_R}) f_{carb}^{N_{age}-1} f_{calc}^{N_{age}}}{((F_0 + F_R) + f_{carb} f_{calc})^{N_{age}}} \tag{2-27}$$

$$f_{carb} = \frac{X_{carb} - X_{calc}}{X_{ave} - X_{calc}} \tag{2-28}$$

$$f_{calc} = \frac{X_{carb} - X_{calc}}{X_{carb}} \tag{2-29}$$

Once the geometric series for X_{ave} (Eq. 2-25) is solved, the average activity of the sorbent can be calculated as function of process (F_0, F_R, f_{carb}, f_{calc}) and sorbent related parameters (a_1, a_2, b, f_1, f_2), respectively [183]:

$$X_{ave} = (F_0 + F_R X_{ave})f_{calc}\left[\frac{a_1 f_1^2}{F_0 + F_R f_{carb} f_{calc}(1 - f_1)} + \frac{a_2 f_1^2}{F_0 + F_R f_{carb} f_{calc}(1 - f_2)} + \frac{b}{F_0}\right] \qquad (2\text{-}30)$$

2.5.6.2 Carbonator Modelling

Several carbonator models have been published for the calculation of the CO_2 absorption in the carbonator. The models might be distinguished according to the assumptions made in the course of the mathematical formulation of hydrodynamics, sorbent deactivation and kinetic approach. This section provides an overview of relevant models with the associated key characteristics [183].

One of the earliest carbonator models from Shimizu et al. (1999) [107] is based on the first order kinetic model of Eq. 2-31:

$$\frac{dX_{carb}}{dt} = c_{CO2,ave} k(X_N - X) \qquad (2\text{-}31)$$

Here, k is the reaction rate constant, $c_{CO2,ave}$ represents the average CO_2 concentration in the carbonator, X_N is the maximum sorbent conversion regressed from 4 complete carbonation-calcination cycles. The hydrodynamic estimates is based on a bubbling fluidized bed (BFB) reactor system that takes into account a two-region model from Kunii and Levenspiel [83].

The first carbonator model that takes into account a function for the sorbent deactivation dependent on related CaL process conditions was published by Abanades et al. in 2004 [184]. In this case, a two-region BFB is considered for the estimation of carbonator hydrodynamics. The deactivation correlation of Eq. 2-18 is combined with a spherical grain model for the calculation of the reaction kinetics:

$$\frac{dX_{carb}}{dt} = \begin{cases} k_X X_N (1 - X_{carb})^{\frac{2}{3}} (c_{CO2} - c_{CO2,eq}), & for X < X_N \\ 0, & for X > X_N \end{cases} \qquad (2\text{-}32)$$

Here, the carbonation reaction rate is set to zero, once the sorbent conversion at the end of the fast reaction regime, X_N, has been reached. The reaction rate in the fast regime is determined as a function of the difference of the CO_2 concentration, the maximum sorbent conversion at the end of the fast regime, the actual carbonation degree and the reaction rate constant, k_X, which is based on the expression of Bhatia and Perlmutter [185]:

$$k_x = \frac{k_s S_0}{(1 - \varepsilon_0)} \qquad (2\text{-}33)$$

In the above equation, k_s is the initial carbonation reaction rate, S_0 is the initial specific surface area of Ca-particles and, ε_0 is the initial particle porosity. The results of this model were validated by results from a batch BFB carbonator.

A CFB carbonator is assumed by the model from Hawthorne et al. (2008) [186]. The solid distribution is calculated by a bubble and emulsion phase model of Kunii and Levenspiel [83] for the lower region of the riser and a core-annulus model of Pugsley and Berutti [187] for the upper part of the riser. The sorbent deactivation is calculated as follows:

$$X_{ave} = \sum_{N=1}^{N=\infty} \frac{F_0}{F_R}\left(1 - \frac{F_0}{F_R}\right)^{N-1} X_{N,pc} \tag{2-34}$$

Here, the evaluation of the sorbent activity is determined as a function of solid population and the maximum sorbent activity as result of partial carbonation, $X_{N,pc}$. The kinetic of the carbonation reaction is calculated according to Eq. 2-35:

$$\frac{dX_{carb}}{dt} = \frac{k_S S_0}{(1 - \varepsilon_0)}(X_{ave} - X_{carb})^{\frac{2}{3}}(c_{CO2} - c_{CO2,eq.}) \tag{2-35}$$

The model of Lasheras et al. (2011) [188] assumes a core-annulus carbonator with a lean region in the bottom part of the riser and the upper dense region based on the model from Kunii and Levenspiel [189]. The sorbent deactivation is calculated by Eq. 2-18. Furthermore, the kinetic model of Eq. 2-32 and 2-33 is applied. The results of this model were validated with reasonable accordance by experimental data of the 1 MW$_{th}$ CaL pilot plant at Technische Universität Darmstadt [190]. Furthermore, the process model has been applied for the design and layout of a 20 MW$_{th}$ CaL demonstration plant [191, 192].

Ylätalo et al. (2012) [193] proposed a CFB carbonator model that applies a vertical density profile based on the correlation of Johnsson and Leckner [194]. In contrast to the aforementioned carbonator models, the temperature profile of the carbonator is derived from an energy balance for each control volume. A sorbent deactivation correlation is not considered, instead an average maximum conversion of 0.134 is assumed. The reaction kinetics are derived as follows:

$$\frac{dX_{carb}}{dt} = m_s(X_N - X_{carb})k_{carb}(c_{CO2} - c_{CO2,eq}) \tag{2-36}$$

Here, the reaction rate is derived as function of the initial carbonation reaction rate, k_{carb}, the difference of the average and the equilibrium partial pressure of CO_2 and the total mass of solids in the system, m_S.

In the carbonator model of Cormos and Simon (2015, 2013) [195, 196], the solids distribution of the CFB carbonator is derived from the approach of Kunii and Levenspiel [189] assuming a lean region and a dense region. The temperature distribution is calculated by an energy balance. Sorbent deactivation is derived from Eq. 2-19, while the kinetics of the carbonation reaction are calculated as follows:

$$\frac{dX_{carb}}{dt} = k_S S_N (1 - X_{carb})^{2/3}(c_{CO2} - c_{CO2,eq}) \tag{2-37}$$

In the above equation, the specific surface area of the sorbent, S_N, in the N-th carbonation-calcination cycle is applied in the course of the kinetic calculation. The accuracy of this model has been verified by means of experimental data from a lab-scale carbonator at INCAR-CSIC.

The CaL process model of Atsonios et al. (2015) [197] combines an advanced thermodynamic process model with a CFD analysis. More precisely, a constant solid volume fraction is assumed for each of the distinguished hydrodynamic regimes. The results of this attempt were validated by experimental data of a cold flow model and of a pilot scale FB test rig at the Institute for Combustion and Power Plant Technology at the Universität Stuttgart.

The model of Romano (2012) [180] is the first attempt to directly include the influence of sulfur on the decay of the sorbent activity. By now it is considered as one of the most accurate attempts for the modelling of sorbent activity and carbonator CO_2 absorption. Therefore, it is adopted and applied in the course of this thesis throughout Chapter 4 and Chapter 5.

2.5.7   Techno-Economic Carbonate Looping Process Characteristics

Within this chapter, an overview of the techno-economic CaL process characteristic is given. This overview focuses on the application of the CaL process at power plants for the delivery of electricity only. The following principles are applicable to coal, natural gas, and biomass fired power plants as well as to WtE plants. In case of an application in the framework of industrial plants, the evaluation methodology needs to be extended, as additional products/services need to be taken into account. Figure 2-17 shows the main material and heat/power streams for a coal-fired power plant retrofitted by the CaL process.

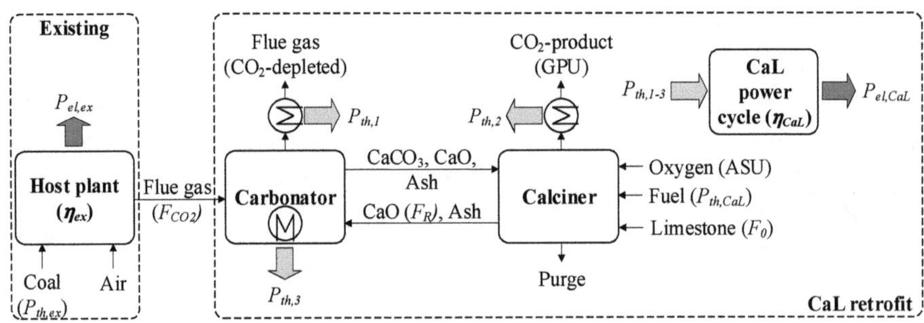

Figure 2-17: Thermodynamic schematic of a CaL retrofit at existing coal-fired power plants.

If a CaL unit is retrofitted to an existing power plant, the heat input and also the electricity output of the total system increases significantly. As the CaL process operates at elevated temperatures (T_{CaL}: 650 - 900 °C), the excess heat is recovered in a dedicated water-steam cycle delivering additional power to the grid. There are three heat sources available for steam generation in the CaL water-steam cycle. The sensible heat of the CO_2-depleted flue gas at the outlet of the carbonator ($P_{th,1}$, $T_{Pth,1}$ = 650 °C) and of the CO_2-product stream at the outlet of the calciner ($P_{th,2}$, $T_{Pth,2}$ = 900 °C) as well as the cooling duty of the carbonator reactor ($P_{th,3}$, $T_{Pth,3}$ = 650 °C). Due to the high temperature of the available heat sources, the operation of a supercritical water-steam cycle is feasible [198, 199].

The CaL process is prone to additional auxiliary power consumption, which results in an inevitable efficiency penalty in contrast to the fuel conversion in a state-of-the-art power plant. Comparing the characteristics and the associated structure of auxiliary power consumption of a CaL unit with a state-of-the-art power plant, the major causes of the efficiency penalty are related to the operation of the ASU and the GPU, the non-recoverable heat consumed by the first calcination of the make-up limestone and the additional solid induced pressure drop of the CFB reactors. Table 2-5 provides an overview of the main thermodynamic characteristics for exemplary full-scale integration studies of the CaL process.

It needs to be noted that each of the cited studies is based on different boundary conditions, such as the type of calculation sequence that is applied for the determination of the required limestone make-up feed to achieve the desired CO_2 absorption efficiency, the operation temperature of the calciner, the water-steam cycle parameters that are applied for the CaL water-steam cycle or the oxyfuel combustion conditions in the calciner. Nevertheless, this table gives the range of crucial parameters of a CaL application in the field of coal-fired power plants.

Table 2-5: Thermodynamic characteristics of CaL retrofit studies at coal-fired power plants.

Parameter	Hawthorne [198]	Ströhle [200]	Martinez [201]	Vorrias [202]	Rolfe [203]	Haaf [4] [192]
F_0/F_{CO2}, -	-	0.03	0.10	0.10	-	0.38
F_R/F_{CO2}, -	7.00	3.00	5.00	-	-	12.6
E_{carb}, -	80.0	80.0	80.0	89.7	80.0	80.0
E_{tot}, %	88.0	87.3	91.5	93.9	90.6	92.9
T_{carb}, °C	650	650	650	650	650	650
T_{calc}, °C	900	900	950	900	900	950
$P_{th,cal}/P_{th,tot}$, %	40.9	37.6	56.9	36.7	48.1	-
$P_{el,net,ex}$, MW$_e$	1052	1052	350	304	620	600
$\eta_{net,ex}$, %	45.6	45.6	36.0[5]	39.1	41.2	-
$\eta_{net,tot}$, %	39.2	42.8	32.6	34.1	33.8	-
$\Delta\eta$, %-points	6.36[6]	2.75[7]	8.30[6]	4.96[6]	7.4[6]	-

All studies assume an operation temperature of 650 °C for the carbonator, while 900 °C are considered as operation temperature for the calciner in almost all cases. Carbonator CO_2 absorption efficiencies between 80 and 89.7 % result in total CO_2 capture efficiencies in the range of 87.3 to 93.9 %. The net electrical efficiency penalty ranges from 4.96 and 7.4 % (including CO_2-compression). The share of the CaL calciner thermal duty among the total thermal duty ranges from 36.7 to 56.9 %. Here, a small number is feasible as it reduces the size of the CaL process with respect to the upstream host plant.

The appropriate evaluation methodology for the net electric efficiency drop is of concern as a retrofit of the CaL process basically implies the installation of a new power plant with current state-of-the-art technology. Especially for retrofit studies that deal with relatively old host plants (i.e. poor net electrical efficiency), the application of a nominalized reference net electrical efficiency, $\eta_{net,ref}$, according to Eq. 2-7 is crucial [183]:

$$\eta_{net,ref} = \frac{P_{th,ex}}{P_{th,ex} + P_{th,CaL}}\eta_{net,ex} + \frac{P_{th,CaL}}{P_{th,ex} + P_{th,CaL}}\eta_{net,CaL} \tag{2-38}$$

Here, $\eta_{net,ex}$ is the net electrical efficiency of the host plant, and $\eta_{net,CaL}$ is the net electrical efficiency of the CaL unit. Both parameter are weighted by the corresponding share of the thermal duty as feed to the corresponding system. Figure 2-18 illustrates the need for the introduction of a reference net electric efficiency by two exemplary CaL process integration cases. Here, the host plant and the CaL process rely on the same type of fuel.

---

[4] The framework of this study is limited to an assessment of the CaL solid looping cycling, excluding the modelling of heat recovery systems.
[5] The net electrical efficiency of the hypothetical power plant is 45 %.
[6] w/ CO_2-compression auxiliary load
[7] w/o CO_2-compression auxiliary load

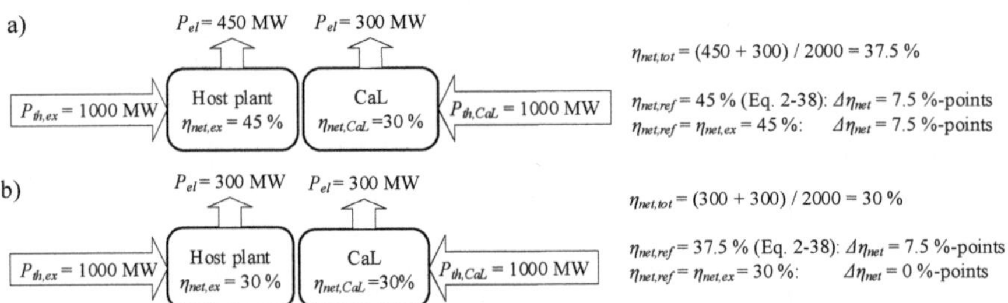

Figure 2-18: Evaluation methodology for two different CaL process integration cases, adopted from [183].

In case (a) the host plant to be equipped with a CaL unit is a state-of-the-art water-steam cycle for power generation. Both evaluation approaches thus give the same value for the net electrical efficiency drop (7.5 %-points in this case). If the electric efficiency of the host plant is rather low (case b), the efficiency penalties are different, depending on the reference efficiency that is applied. If the electric efficiency of the host plant is applied, the efficiency penalties become zero due to the rather efficient heat utilization in the water-steam cycle of the CaL process. Therefore, the reference electric efficiency needs to be calculated according to Eq. 2-38. Doing so, the efficiency penalties are determined to be 7.5 %-points. In the above cited studies, this methodology was applied in the course of the work of Martinez et al. [201], while the integration study of Vorrias et al. [202] rely on the net electrical efficiency of the host plant as reference, which results in quite a moderate net electrical efficiency penalty.

The economics of CCS in the framework of power generation systems are typically discussed by means of the levelized cost of electricity (LCOE) and the cost of CO_2 avoidance (CAC) [204, 205]. The LCOE measures the unit cost of a power plant with and without CO_2 capture, and approximates the average discounted electricity price over the project duration that would be required as an income to match the net present value of costs of the project. The CAC approximates the average discounted CO_2 credit (tax or quota) over the duration of the project that would be required as an income to match the net present value of costs due to the CO_2 capture implementation. A detailed mathematical definition of these parameters is given in Chapter 5.2.3. Table 2-6 summarizes the economic characteristics of exemplary full-scale CaL retrofit studies.

Table 2-6: Economic characteristics of CaL retrofit studies at coal-fired power plants.

Parameter	Rolfe et al. [206]	Cormos [200]	Yang et al. [207]
$LCOE_{Ref}$, EUR/MWh$_e$	60.9	45.5	35
Cost reference year, -	2018	2014	2010
$LCOE_{CaL}$, EUR/MWh$_e$	77.3	68.4	54.3
CAC, EUR/t$_{CO2,av}$	22.3	31.3	28.9

The LCOE of a CaL process retrofitted power plant increase by 27 - 55 % compared to the reference state. The derived CAC vary between 22 and 31 EUR/t$_{CO2,av}$. Thus, among other processes for CO_2 capture such as oxyfuel combustion (33 - 60 EUR/t$_{CO2,av}$) [208, 209] MEA-scrubbing (56 - 79 EUR/t$_{CO2,av}$) [210, 211], membranes (45 - 72 EUR/t$_{CO2,av}$) [212, 213] or the IGCC process (42 - 87 EUR/t$_{CO2,av}$) [211, 214], the CaL process offers substantial improvements regarding the economic performance.

3 Experimental Investigations at 1 MW$_{th}$ Scale

This chapter covers the experimental investigations at the 1 MW$_{th}$ CaL pilot plant. Based on a detailed description of the 1 MW$_{th}$ CaL pilot plant, the CO$_2$ capture performance of the CaL process in the framework of a WtE plant is assessed. This particularly implies the utilization of SRF as supplementary fuel in the calciner and the decarbonization of flue gas with a composition similar to that of a WtE plant fueled by MSW. The evaluation is based on the online measurement data from the process control system and on the results from the solid sample analysis, respectively. A performance assessment is carried out for each of the CFB reactors, e.g. carbonator and calciner as well as for the comprehensive CaL process. Furthermore, the effect of various process conditions on the chemical composition and properties of the sorbent are analyzed. The evaluation of the calciner covers an assessment of the emissions of major gaseous pollutants and the assessment of sorbent regeneration.

The experimental investigations were carried out in 2018 in two consecutive test campaigns, each lasting for approximately 14 days. For the first time worldwide, a waste-derived fuel was utilized as supplementary fuel in the calciner of the CaL process. Overall, the CaL pilot plant was operated more than 230 hours under representative CaL operation conditions.

3.1 Carbonate Looping Process Configuration

The following section describes the setup of the 1 MW$_{th}$ CaL pilot plant. In order to enable the utilization of SRF in the oxyfuel calciner, a new fuel feeding system was designed, installed and commissioned. Figure 3-1 shows a schematic of the 1 MW$_{th}$ CaL pilot plant. Related to the flowsheet, Figure 3-2 presents a 3D-CAD sketch of the 1 MW$_{th}$ CaL pilot plant including relevant auxiliary systems. The CaL pilot plant consists of three main subsystems. The dimension of the main subsystems are summarized in Table 3-1. In addition to the CFB carbonator and the CFB calciner, the on-site combustion chamber allows for the investigation of the CO$_2$ capture performance based on pulverized coal or natural gas-originated flue gas, respectively. Additionally, there is the possibility to introduce technically pure CO$_2$ into the primary air of the carbonator, which in fact allows investigations based on a synthetic flue gas stream.

Table 3-1: Dimensions of the main subsystems of the 1 MW$_{th}$ CaL pilot plant.

System	d_i (m)	d_o (m)	h (m)
Combustion Chamber	1.00	1.30	7.00
CFB Carbonator	0.59	1.30	8.66
CFB Calciner	0.40[8]	1.00	11.4

In the carbonator, CO$_2$ is absorbed from the flue gas stream. The primary gas is then fed to the carbonator riser through the nozzle grid at the bottom of the reactor. The inlet temperature of the primary gas stream can be controlled by means of electrical air preheaters up to approximately

---

[8] The calciner is conical in the bottom region. From 0.28 m at a height of 0 m the diameter widens until 0.40 m at a height of 1.15 m.

350 °C. The CO_2-lean flue gas at the carbonator outlet is released into the environment after heat removal and particle cleanup downstream of the carbonator cyclone. The partially carbonated solids are separated in the carbonator cyclone and recirculated back to the carbonator riser, or transported to the calciner. The temperature of the carbonator can be adjusted by means of five axially arranged cooling tubes, attached on the top of the reactor. The amount of heat being removed is adjusted by the immersion depth of the cooling tubes inside the carbonator riser. Carbonator bottom ash is discharged by a water cooled screw conveyor, which is mounted on the extraction pipe below the carbonator nozzle grid.

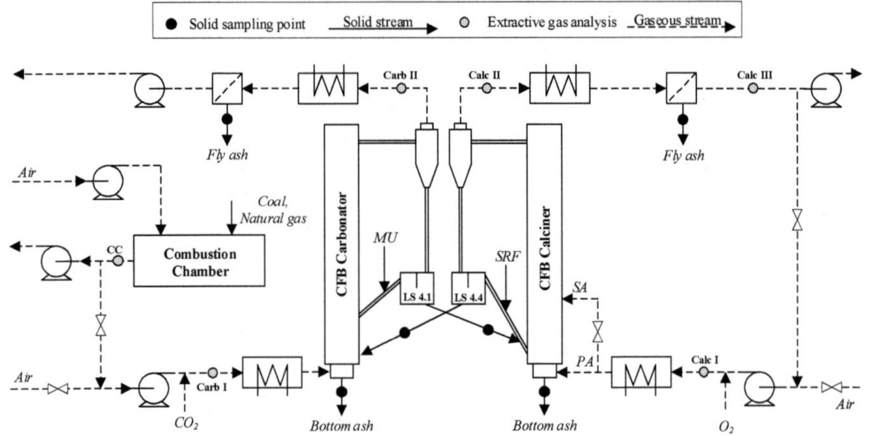

Figure 3-1: Flowsheet of the 1 MW$_{th}$ CaL pilot plant at Technische Universität Darmstadt.

The solid transfer between carbonator and calciner is carried either out by means of a mechanically controlled screw conveyor or a non-mechanically transport valve, the so-called "L-Valve" [215]. The screw conveyor offers a nearly linear material conveying behavior and a smooth process controllability, whereas this device is prone to additional maintenance because of the moving parts and sealing requirements. Additionally, heat is removed from the circulating sorbent due to the cooling system inside the conveying screw, which leads to a higher specific heat requirement in the calciner. The "L-valve" is relatively wear-resistant, due to its simple construction. The solids carryover is realized by a constant flux of a fluidization agent in the horizontal part. However, the controllability of the solid mass flow is rather multi-layered and complex. The solid carryover depends on the local hydrodynamic conditions, which are further influenced by the solid inventories of carbonator and calciner. Thus, a change in the fluidization rate in the L-valve might lead to varying effects among different solid loads of carbonator and calciner.

The partly carbonated solid stream is fed to the calciner return leg. At this point, the globally circulating solid stream is mixed together with the internally circulating solid stream of the calciner and the SRF. This mixture is then introduced to the riser of the calciner. In the riser, the incoming solids are instantaneously heated and calcined. The combustion air of the calciner consist of technically pure oxygen ($y_{O2} > 99$ vol.%) and recirculated off-gas that is extracted downstream of the calciner particle filter. The temperature of the combustion air can be raised up to approximately 450 °C by means of an electrical preheating system. After being preheated, part of the combustion air is fed to the riser via the calciner nozzle grid. The other part is introduced to the riser secondary air joints at a height of 2.9 m above the nozzle grid. The CO_2-rich calciner off-gas undergoes heat removal and particle cleanup. Thereafter, a slip stream is recirculated to the inlet of the primary air for oxygen dilution, the other part is released to the environment. The regenerated solid stream that is separated in the calciner cyclone is partly recirculated back to the riser, whereas the other part is fed to the carbonator. The split ratio of the solid phase is controlled by means of a cone valve attached to the loop seal of the calciner. Bottom ash is discharged from the calciner via a water cooled screw conveyor attached to the extraction pipe at the calciner nozzle grid.

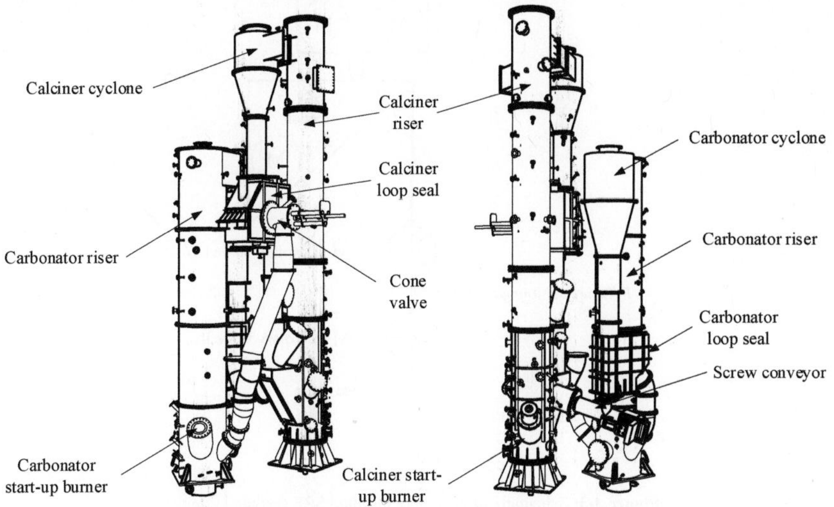

Figure 3-2: 3D-CAD drawing of the 1 MWth CaL pilot plant.

Both CFB reactors are fully refractory lined in order to minimize heat losses. Furthermore, the major components of the pilot plant are similar or equal to industrial standard, i.e. a propane-fired start-up burner for each CFB reactor, two-pass heat exchangers, and bag filters. The cooling system of the pilot plant consists of a pressurized cooling-water closed loop that operates at 16 bar_{abs} and 120 °C. Heat is released to the environment via air cooled heat exchangers. The cooling of the screw conveyor is achieved by a separated cooling-oil based closed loop, operating at 2 bar_{abs} and 160 °C. Again, heat is released to the environment via air cooled heat exchangers.

The limestone make-up is fed directly into the return leg of the carbonator. The dosing system for limestone comprises a supply vessel and an underneath attached gravimetric dosing vessel. The desired mass flow of make-up limestone is adjusted by the rotating speed of the screw feeder.

The SRF feeding system was newly designed, installed and commissioned ahead of the experimental investigations. Figure 3-3 shows a schematic of the SRF feeding system. It consists of a supply bunker, a subsequently arranged gravimetric dosing system, and a system of two rotary valves at the interface to the hot process parts of the calciner. The supply bunker is manually filled with SRF fluff from bigbags. It holds up a supply volume of nearly 2 m³. The gravimetric dosing system is based on the principle of a controllable upwards moving belt, covered by a returning wheel on the upper end. The SRF mass flow is controlled by the velocity of the belt and the free cross sectional area at the dropping point of the dosing system. Once the dosing system runs empty, the supply bunker automatically starts refilling the feeder hopper. In order to avoid any leakage of process gases to the environment, two consecutive arranged rotary valves are attached below the dosing system at the process interface. The free space between the two rotary valves is constantly flushed by pressure sealing gas in order to achieve a pressure gradient towards the process. For a stable transport behavior of SRF, a constant flow of transport gas is fed to the process below the second rotary valve.

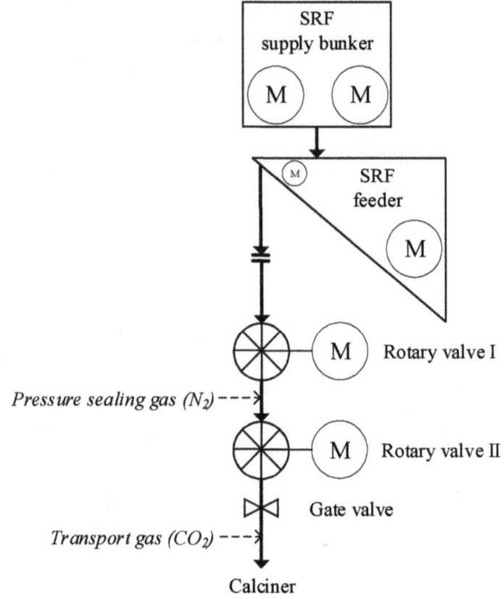

Figure 3-3: Schematic of the newly installed SRF feeding system.

3.2 Measurement Equipment

The 1 MW$_{th}$ pilot plant is equipped with numerous measurement devices at relevant process positions. The safe operation of the pilot plant is guaranteed by continuous data logging. All measured values are automatically saved on the process control system each second. A nominal operation range is defined for each parameter. Once a given threshold is passed, an alarm appears on the process control screen. Most important for the process evaluation are the values for pressure, temperature, volumetric flow rates, mass flow rate and gas composition. This chapter provides an overview of the applied measurement equipment and the corresponding accuracy of each measurement.

3.2.1 Pressure

A main application of the pressure measurement devices are the riser of the CFB reactors (e.g. carbonator, calciner) and the coupling components (e.g. loop seals). Based on the solid induced pressure drop, important hydrodynamic parameters such as the solid inventory or the void fraction of the fluidized bed can be derived. Additionally, knowledge of pressure drop and absolute pressure is required for the calculation of the flow rate via various flow meters, such as venture or orifice plate flow measurement. In addition to the pressure measurement devices for process evaluation, there are various measurement devices required to allow a safe operation of the pilot plant. For example, it is important to note the pressure within the cooling systems or the pressure drop within the flue gas ducts.

Different types of pressure measurement devices are applied at the 1 MW$_{th}$ pilot plant. The pressure in the CFB reactors and in the solid looping system is measured relative to the environment. In contrast, the pressure in the cooling system, in the distribution network of technical gases and in the flue gas and primary air ducts is measured by a difference pressure indicator that allow for a higher accuracy.

A summary of the pressure measurement position related to the solid looping systems is given by Table 3-2. As reference height, the nozzle grid of each CFB reactor riser is considered. Especially in the bottom region, of each riser, there are several pressure measurement devices installed to allow for a high-resolution measurement in the dense region of the fluidized bed. The applied pressure measurement devices for the reactor system show a measuring inaccuracy < 0.5 %$_{Range}$, whereas the a measuring inaccuracy < 0.1 %$_{Range}$ is achieved by the pressure transductors associated with volumetric flow measurements.

Table 3-2: Position of pressure measurement devices within the solid looping system.

System	Location	Height above reactor riser nozzle grid (m)
Carbonator	Reactor riser	0.00, 0.10, 0.22, 0.40, 0.58, 0.91, 1.10, 2.07, 3.40, 8.02, 8.08
	Loop seal	2.84, 3.83, 4.45, 6,01
Calciner	Reactor	0.00, 0.09, 0.25, 0.41, 0.62, 1.65, 2.15, 3.39, 4.29, 5.50, 8.13, 9.81
	Loop Seal	5.72, 6.57, 7.07, 8.15

3.2.2 Temperature

Knowledge of the temperature at certain process locations is crucial from a safety and process evaluation point of view. The temperature distribution within the CFB reactor risers allows for the assessment of carbonation and calcination phenomenon. Furthermore, the control of the oxyfuel combustion in the calciner is related to the temperature profile in the riser. In the supply network of technical gases, knowledge of the temperature is required for the calculation of the mass flows of the gases. From a safety point of view, all rotary valves were equipped with temperature measurement devices in order to directly detect the leakage of hot process gases.

Inside the CFB risers, the temperature measurements consist of type-N transducers that are covered with a high-temperature erosion resistant metallic protection tube. The measurement range is from -200 up to 1200 °C with a maximum absolute error of 2.5 K. A summary of the pressure measurement position of the solid looping system is given by Table 3-3.

Table 3-3: Position of temperature measurement devices within the solid looping system.

System	Location	Height above reactor riser nozzle grid (m)
Carbonator	Reactor	0.25, 0.29, 0.93, 1.55, 1.55, 1.68, 2.67, 6.96, 7.75, 8.21
	Loop seal	2.31, 2.78
Calciner	Reactor	0.28, 0.49, 1.28, 1.78, 3.38, 7.04, 9.73
	Loop Seal	2.67, 6.46

3.2.3 Gas Composition

Extractive gas measurement devices are used exclusively at the pilot plant. The process gases are extracted by a sampling probe, transported to the measurement device via a heated gas tube, upgraded and cleaned, and finally introduced to the gas analysis device. All gas analyses, except the FTIR measurement, rely on a measurement of dry process gases. The derived gas composition thus needs to be corrected for the moisture content of the corresponding gas stream. A summary of the applied gas measurement techniques is given by Table 3-4. Each extraction point is additionally marked in Figure 3-1.

Table 3-4: Gas measurement techniques at the 1 MW_{th} CaL pilot plant.

Extraction point	Species	Technique	Range	Inaccuracy ($\%_{Range}$)
Combustion chamber (CC)	O_2	Paramagnetic	0 - 100 vol.%	< 0.5
	CO_2	NDIR[9]	0 - 30 vol.%	< 0.5
	CO	NDIR	0 - 5 vol.%	< 0.5
	SO_2	NDIR	0 - 4000 ppm_v	< 0.5
	NO	NDIR	0 - 1000 ppm_v	< 0.5
	H_2O	Psychometric	2 - 100 vol.%	< 1
Carbonator inlet (Carb I)	O_2	Paramagnetic	0 - 100 vol.%	< 0.5
	CO_2	NDIR	0 - 100 vol.%	< 0.5
Carbonator outlet (Carb II)	O_2	Paramagnetic	0 - 100 vol.%	< 0.5
	CO_2	NDIR	0 - 30 vol.%	< 0.5
	CO	NDIR	0 - 5 vol.%	< 0.5
	SO_2	NDIR	0 - 4000 ppm_v	< 0.5
	NO	NDIR	0 - 1000 ppm_v	< 0.5
Calciner inlet (Calc I)	O_2	Paramagnetic	0 - 100 vol.%	< 0.5
	CO_2	NDIR	0 - 100 vol.%	< 0.5
Calciner outlet (Calc II)	O_2	Paramagnetic	0 - 25 vol.%	< 0.5
	CO_2	NDIR	0 - 40 vol.%	< 0.5
	CO	NDIR	0 - 40 vol.%	< 0.5
	SO_2	NDIR	0 - 5 vol.%	< 0.5
	NO	NDIR	0 - 1000 ppm_v	< 0.5
	HCl	FTIR[10]	0 - 1000 ppm_v	< 1
Calciner off-gas (Calc III)	H_2O	Psychometric	2 - 100 vol.%	< 1

The composition of the gas streams in the CaL process is required for the calculation of the gas phase properties (e.g. density, dynamic viscosity) that are further applied in the course of volumetric gas flow calculation based on a given pressure drop (e.g. venture, orifice plate). Furthermore, it determines the flux of the gas species that are relevant for process evaluation (e.g. CO_2, O_2, N_2 and H_2O).

3.2.4 Volumetric Gas Flow

The volumetric flow rate of relevant process gas streams is either derived from an orifice plate or from a venturi flow measurement. In both cases, the composition, the temperature and the absolute pressure at the point of the flow measurement need to be known to achieve the given accuracy. In some cases, the gas composition is nearly constant, for instance if ambient air is utilized as fluidization agent. In other cases, the gas composition is a function of process performance (i.e. the flue gas streams of carbonator and calciner). Based on the pressure drop induced by the flow of the gaseous

---

[9] Non-dispersive infrared
[10] Fourier-transform-infrared-spectrometer

medium, the flow quantity is derived and logged by the process control system. A summary of the applied volumetric flow measurements at the 1 MW$_{th}$ pilot plant is given by Table 3-5.

Table 3-5: Volumetric flow measurement techniques at the 1 MW$_{th}$ CaL pilot plant.

Process position	Medium	Technique	Range (Nm³/h)	Inaccuracy (%$_{oRange}$)
Carbonator (inlet)	Flue gas / Air	Ring chamber orifice plate	130 - 1,300	< 1.0
	CO$_2$	Standard orifice plate	18 -180	< 1.0
Carbonator (outlet)	Flue gas	Venturi flow meter	125 - 1,250	< 4.0
Calciner (inlet)	Recirculation gas	Venturi flow meter	100 - 1,000	< 4.0
	O$_2$	Standard orifice plate	22 - 222	< 1.0
Calciner (outlet)	Flue gas	Venturi flow meter	130 - 1,300	< 4.0
Combustion Chamber (inlet)	Air[11]	Ring chamber orifice plate	76 - 760	< 1.0
	Natural gas	Mass flow controller	3.2 - 16	< 1.5
Combustion chamber (outlet)	Flue gas	Venturi flow meter	120 - 1,200	< 4.0
Loop seals	Air	Ring chamber orifice plate	5 - 50	< 1.0

3.2.5 Solid Mass Flow

The solid mass flow of the make-up limestone and of the SRF is indirectly measured by means of gravimetric measurement devices. Based on the mass change of the corresponding dosing vessel in a certain time interval, the mass flow is automatically derived by the process control system. The mass change is determined by four measurement cells below each vessel. Each dosing vessel needs to be mechanically decoupled from the steel structure to allow for a precise measurement. The inaccuracy of both gravimetric measurement devices is approximately < 1.0 %$_{Range}$.

3.2.6 Accuracy of Process Evaluation

In the course of the experimental investigation, several directly or indirectly measured parameters were combined. Directly measured parameters are for instance pressure, temperature or the volumetric composition of a gas flow, whereas indirectly measured parameter are calculated based on more than one directly measured parameter. This is for instance the volumetric gas flow which is derived from the corresponding gas composition, its temperature and pressure as well as the flow induced pressure drop over a given geometry.

In case of a directly measured parameter, the relative error given by the supplier of the measurement device is applicable for the estimation of the uncertainty, while in case of indirectly measured parameter, all relative errors of the input variables need to be taken into account.

In case of normally distributed uncertainties, the Gaussian error propagation approach is typically applied. For example, the relative measurement uncertainty of the flue gas volumetric flow rate yields to approximately 1.0 - 1.5 %. It needs to be noted that this represents a relatively simple indirectly measured parameter, whereby seven directly measured parameters are considered. In case of the CO$_2$ absorption efficiency of the carbonator (E_{carb}) or the total CO$_2$ capture efficiency (E_{tot}), the complexity of the calculation of the uncertainty is significantly increased as several directly and indirectly

---

[11] Here, only the measurement of on airflow is given representative for the different types of air supply, e.g. auxiliary-, axial-, swirl- and staging air.

measured parameters are combined. Therefore, in accordance with previous CaL process evaluation attempts at the 1 MW$_{th}$ pilot plant, the uncertainties of all relevant process evaluation parameters are assumed to be less than 5.0 % [216, 217].

3.3   Applied Materials

In the CaL test campaigns, a natural limestone from a German limestone quarry was used as sorbent feed. Table 3-6 summarizes the chemical composition of the limestone.

Table 3-6: Chemical composition of the fresh make-up limestone.

Component	Chemical formula	Mass fraction (wt.%)
Calcium carbonate	$CaCO_3$	98.2
Magnesium carbonate	$MgCO_3$	1.20
Silicon dioxide	SiO_2	0.40
Sulphur trioxide	SO_3	< 0.01
Iron (III) oxide	Fe_2O_3	< 0.10
Aluminum oxide	Al_2O_3	0.10

The particle diameter of the limestone ranges from 25 to 500 µm with a median diameter of 179 µm. The particle size distribution of the limestone is given by Figure 3-4.

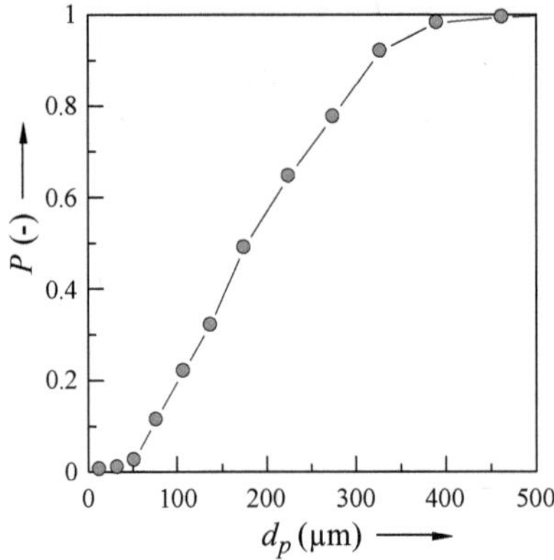

Figure 3-4: Particle size distribution of the fresh limestone make-up.

Two different types of industrial SRF were used during the experimental investigations. The SRF was processed from non-recyclable industrial waste, bulky wastes, commercial wastes and residues from sorting plants [218]. Table 3-7 summarizes the proximate analysis, elementary composition and the lower heating value of the two types of SRF. Both types were fed to the CaL process without any particular pre-treatment in the form of raw fluff.

The numbers given in the table represent the average values for the SRF in the given time frame of the experimental investigation in 2018. It needs to be noted that within each delivery, the SRF composition and properties are likely to vary noteworthy as a consequence of fluctuations in the feed streams to the SRF process units. Nevertheless, a detailed statistically evaluation of SRF composition was beyond the scope of this thesis.

Table 3-7: Proximate analysis, LHV and elementary composition of the two types of SRF.

Component	Unit	SRF I	SRF II
Moisture	wt.%$_{ar}$	12.3	19.4
Ash	wt.%$_{ar}$	9.70	15.4
Volatiles	wt.%$_{ar}$	71.2	56.7
Fixed carbon	wt.%$_{ar}$	6.84	8.50
LHV	MJ/kg	21.4	15.6
C	wt.%$_{db}$	58.0	47.2
H	wt.%$_{db}$	8.20	6.50
N	wt.%$_{db}$	0.54	1.20
S	wt.%$_{db}$	0.22	0.36
Cl	wt.%$_{db}$	0.73	0.91
O	wt.%$_{db}$	21.3	24.7

Both types of SRF are characterized by a relatively large share of volatile matter, whereas the fixed carbon represents a minor share of the total combustible matter (< 10 %). This is caused by the quantities of plastics, biomass and paper/cardboard that are present in the feed streams to the SRF processing units. The size specification of the SRF is also of concern for the operability of the pilot plant. The maximum particle size of SRF I was below 30 mm, whereas SRF II needs to fulfill the requirement of being smaller than 50 mm. This is of particular interest when taking into account the various incombustible species in the ashes (e.g. metals, glass and mineral matter). Since the accumulation of larger ash fraction might lead to the disturbances in the hydrodynamics of the interconnected CFB system (see Chapter 3.7.2).

The on-site combustion chamber for the supply of a real flue gas was fueled either by natural gas (NG) or lignite (lignite energy pulverized, LEP). Table 3-8 shows the elementary composition of LEP and NG.

Table 3-8: Elementary composition of the fuels used in the combustion chamber [219, 220].

Species$_{LEP}$	Unit	LEP	Species$_{NG}$	Unit	NG
C	wt.%$_{ar}$	59.9	CH_4	vol.%	92.6
H	wt.%$_{ar}$	4.76	C_2H_6	vol.%	4.77
N	wt.%$_{ar}$	0.69	C_3H_8	vol.%	0.36
S	wt.%$_{ar}$	0.32	CO_2	vol.%	1.42
O	wt.%$_{ar}$	20.7	N_2	vol.%	0.83
Cl	wt.%$_{ar}$	~ 0	S	vol.%	~ 0
Ash	wt.%$_{ar}$	3.80			
H_2O	wt.%$_{ar}$	9.90			

In addition to the real flue gases established by the combustion of NG and LEP, the CaL process was partly supplied by CO_2 enriched ambient air (i.e. synthetic flue gas). This was required during the startup phase of pilot plant operation or in the case of a combustion chamber malfunction. The common characteristic of all flue gas options was the CO_2 concentration similar to typical values of a WtE plant fueled by MSW ($y_{CO2,WtE}$ ~ 9.5 - 10 vol.%).

Table 3-9 summarizes a representative volumetric composition of the flue gases during the course of the experiment depending on the type of flue gas source. Additionally, the typical flue gas composition of a WtE plant is given.

Table 3-9: Flue gas composition for various flue gas options and for a typical WtE plant [24].

Species	Unit	FG I: Synth. flue gas	FG II: NG	FG III: LEP	WtE plant
CO_2	vol.%	9.60	9.76	9.71	10.0
H_2O	vol.%	~ 0	10.9	9.22	11.4
N_2	vol.%	71.5	70.6	73.3	70.3
O_2	vol.%	18.9	8.65	7.73	8.20
SO_2	ppm$_v$	~ 0	~ 3.00	44.6	19.0

The concentration of CO_2 was in a similar range for all cases (y_{CO2}: 9.60 - 9.76 vol.%), which is also due for the concentration of N_2 (y_{N2}: 70.6 - 73.3 vol.%). It needs to be noted that flue gas type FG I represents a dry flue gas ($y_{H2O} \sim 0$ vol.%), in contrast to the flue gas types FG II and FG III (y_{H2O}: 9.22 - 10.9 vol.%). Accordingly, the CO_2 absorption in the carbonator was not positively affected due to the effect of water vapor in the carbonator flue gas (see Chapter 2.5.4). Furthermore, the concentration of SO_2 in case of FG III is relevant as it favors sorbent deactivation. The differences in the oxygen concentration are of less importance with regard to the expected performance of the CaL process.

3.4 Evaluation Methodology

This chapter describes the applied methodology for the evaluation of the experimental data. A distinction is made between the methods for the analysis associated with the solid samples and the evaluation methodology of the online process data.

3.4.1 Solid Sample Evaluation

During the experimental investigations, numerous solid samples were extracted from the pilot plant at crucial process locations. These are for instance the bottom and filter ashes of both CFB reactors as well as samples of the circulating solid material that were extracted from each CFB reactor loop seal. The enclosure of the solid sample analysis results in combination with the online measurement data is essential for a comprehensive evaluation of process characteristics.

Ahead of each sorbent analysis, the solid samples were sieved (mesh width: 1,000 µm) in order to extract and weight the fraction of coarse inert materials (glass, metals, mineral matter). This methodology was required to achieve a more homogeneous solid sample and to fulfill the requirement in terms of maximum particle size for the analysis devices.

3.4.1.1 Chemical Composition

The composition of the solid samples were determined in the laboratory of the Lhoist Business Innovation Center of the Rheinkalk GmbH by various principles. The elementary composition was determined by means of X-Ray fluorescence (XRF) analysis [221]. The mass fraction of chloride in the solid was determined by potentiometric titration [222].

3.4.1.2 Particle Size Distribution

The particle size distribution (PSD) was determined by a sieve analysis following the standards DIN 66165-1 and DIN 66165-2 [223, 224]. Accordingly, the sample of the solid is separated by sieves with increasing mesh sizes and the respective mass fraction is documented. This process represents a machine sieving with a stationary single sieve in a gaseous, moving fluid.

3.4.1.3 Abrasion Coefficient

The particle abrasion coefficient was determined in a test rig proposed by Rýden et al. [225]. The working principle of this test rig implies the exposure of the sorbent to mechanical attrition forces likely to prevail in industrial-scale CFB applications. Thermal induced abrasion effects are not mimicked by this assessment.

The unit is based on a modified version of the cylindrical Davidson beam shell. The fluidization of the particles is achieved by a tangentially arranged inlet nozzle at the bottom of the cone with an inlet velocity of approximately 100 m/s. The particles are then fed into the cone and being sufficiently mixed due to the conical shale of the jet shale. The gravimetric separator is located in the upper part of the test setup and reduces the flow velocity to almost 1/20000 of the nozzle inlet velocity. The coarse particles are being returned to the jet bowl. The very fine particles, which are entrained by the gas flow, are separated by a filter with a pore size of 0.1 µm and collected in a sampling container. The experiments were carried out at room temperature and atmospheric pressure.

Sample preparation is necessary prior to the tests in order to exclude both too coarse and too fine particle fractions from the tests. In this case, the sample was sieved to a particle size of approximately 40 to 355 µm, which is representative to the PSD of the fresh limestone (see Chapter 3.3).

The total cumulative abrasion coefficient, A_{tot}, describes the loss of fine particles over the complete test duration of one hour:

$$A_{tot} = \frac{m_{f,t=60min} - m_{f,t=0min}}{m_S} \cdot 100 \tag{3-1}$$

Here, m_f is the mass of fine particles at the time of experiment, t and, m_S is the initial mass of the sample.

While excluding the first half of the abrasion test, it is possible to calculate the abrasion coefficient, A_i, which is not influenced by initial abrasion effects:

$$A_i = \frac{60}{30} \cdot \frac{m_{f,t=60min} - m_{f,t=30min}}{m_S} \cdot 100 \tag{3-2}$$

3.4.1.4 Sorbent Activity

In order to assess the sorbent activity of a given sorbent sample, a thermogravimetric analysis (TGA) was applied. This implies the measurement of the mass change of a sorbent sample as a function of the prevailing sorbent sample temperature and of the gas atmosphere that the sorbent sample is exposed to. For the investigation of the sorbent activity, the TGA STA449 Jupiter™ from Netzsch was applied [226]. This TGA device consists of a sample holder which is placed in an adjustable furnace. The furnace can be heated up to 1,600 °C with different heating rates. The sample holder is connected to a weighting system. Furthermore, valves and mass flow controllers are applied in order to establish the desired gas atmosphere in the furnace. The sample holder was an alumina type S (d: 17 mm, h: 250 mm) sample carrier system. The sample temperature is adjustable with a resolution of 0.001 K. The resolution of the weighting system is 0.1 µg with an allowable drift of 5 µg/h [226].

In order to establish a carbonation-calcination environment similar to the conditions at the 1 MW$_{th}$ pilot plant, the measurement program, as summarized in Table 3-10, was applied.

Table 3-10: TGA measurement program for the determination of the sorbent activity.

Segment	Temperature (°C)	Heating rate (K/min)	Duration (min)	y_{CO2} (vol.%)	y_{N2} (vol.%)	y_{H2O} (vol.%)
0	25	-	-	-	-	-
1	900	10	88	-	100	-
2	900	-	10	-	100	-
3	650	-10	25	-	100	-
4	650	-	5	-	80	20
5	650	-	20	10	70	20
6	25	-10	63	-	100	-

Ahead of each test run, approximately 20 mg of a sorbent sample was placed on the sample holder. The solid sample is heated up to 900 °C to ensure a complete calcination. Subsequently, the temperature is reduced to 650 °C, and the solid sample is exposed to a synthetic flue gas. After being exposed to carbonation conditions for 20 minutes, the sample is cooled down and the test run is finalized. In the course of the test evaluation, the activity of solid samples was measured from a total of 12 samples each by means of a double determination.

Based on the mass change of the respective limestone sample, the average molar conversion of the limestone sample in the $n+1$ cycle, $X_{ave,n+1}$, can be determined according to Eq. 3-3.

$$X_{ave,n+1} = \frac{m_{CaCO3} - m_{CaO}}{m_{CaO}} \frac{M_{CaO}}{M_{CO2}} \tag{3-3}$$

Here, m_{CaCO3} is the total mass of calcium carbonate, m_{CaO} is the mass of calcium oxide and, M_i are the molar masses of calcium oxide and CO_2 respectively. Based on the resulting carbonation curve, one could derive the maximum average molar carbonation conversion based on a graphical fitting exercise. Figure 3-5 shows the derived carbonation curve for a fresh make-up sample (Graph a) and for a highly cycled sorbent sample extracted from the 1 MW_{th} pilot plant (Graph b).

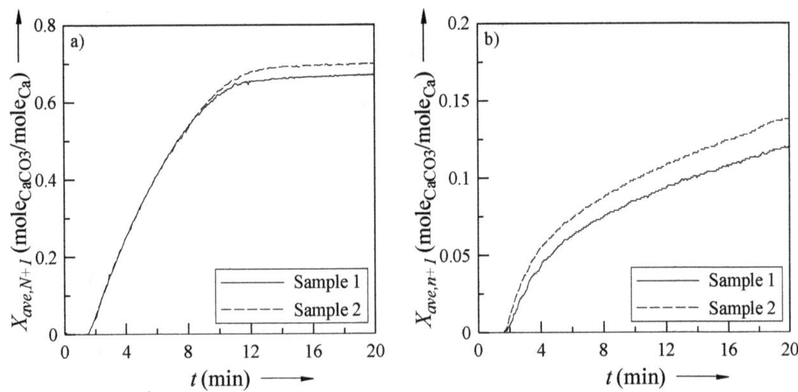

Figure 3-5: TGA analysis of a fresh limestone (a) and of highly cycled sorbent (b).

In both cases, the data derived from the two test runs is depicted. For the fresh make-up sample, the average sorbent activity is 0.670, whereas this parameter is significantly lower in case of the highly cycled sorbent ($X_{max,n+1}$: 0.075). Furthermore, it can be seen that the transition point is less clearly pronounced in the highly cycled sorbent case. This is due to the fact that sample comprises a mixture of particles with a reaction age.

3.4.2   Process Data Evaluation Methodology

The following evaluation methodology was applied in order to discuss the CaL process performance. Figure 3-6 shows a schematic of the CaL process evaluation methodology, including the relevant reactor systems and the corresponding material flows.

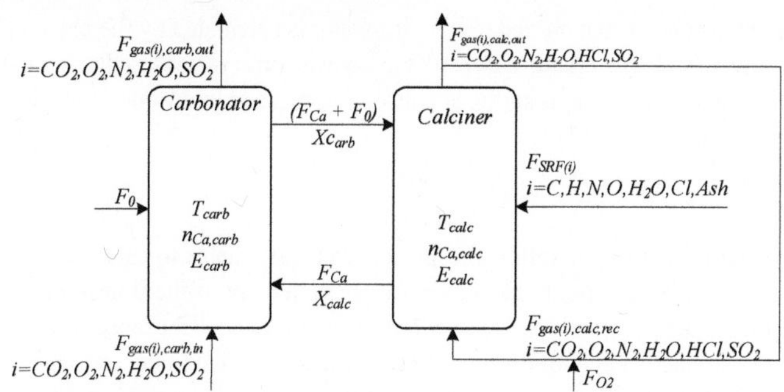

Figure 3-6: Schematic of relevant material streams for CaL process evaluation.

The CO_2 absorption efficiency in the carbonator, E_{carb}, is calculated based on the amount of CO_2 that is absorbed by the solid phase, according to Eq. 3-4.

$$E_{carb} = \frac{F_{CO2} - F_{CO2,carb,out}}{F_{CO2}} = \frac{F_{CO2,abs}}{F_{CO2}} \tag{3-4}$$

Here, F_{CO2} states the molar flow of CO_2 that is introduced into carbonator with the flue gas, $F_{CO2,carb,out}$ states the molar of CO_2 that is present in flue gas after the carbonator. Both quantities are derived from volumetric flow measurement and online gas analysis.

As stated earlier, the theoretical maximum of this parameter, $E_{carb,eq}$, depends on the chemical equilibrium concentration of CO_2 in the gas phase given by the operation temperature of the carbonator and the inlet concentration of CO_2 (see Eq. 2-14 and 2-15). As the inlet concentration of CO_2 was partly altered during the course of the experimental investigations, the nominalized CO_2 absorption efficiency, $E_{carb,nom}$, is applied in the course of the following discussion:

$$E_{carb,nom} = \frac{E_{carb}}{E_{carb,eq}} \tag{3-5}$$

Taking into account the amount of CO_2 released by the combustion of SRF in the calciner, $F_{CO2,SRF}$, and released by the initial calcination of the limestone make-up, F_0, the total CO_2 capture rate of the CaL system, E_{tot}, is calculated according to Eq. 3-6. Due to the inherent characteristics of the oxyfuel calciner, the total CO_2 capture rate always exceeds the CO_2 absorption efficiency in the carbonator.

$$E_{tot} = \frac{F_{CO2,calc,out}}{F_{CO2} + F_{CO2,SRF} + F_0} \tag{3-6}$$

The molar make-up ratio, Λ, is derived from Eq. 3-7 by the ratio of the molar flux of $CaCO_3$ in the fresh make-up limestone and the molar flux of CO_2 contained in the flue gas that is introduced to the carbonator. In this study, the $CaCO_3$ mass fraction in the make-up limestone, which is 98.2 wt.% according to Table 3-6, is assumed to 100 wt.%.

$$\Lambda = \frac{F_0}{F_{CO2}} \tag{3-7}$$

Similarly, the molar circulation rate, Φ, balances the molar flux of Ca-species that are fed to the carbonator, F_R, by the CO_2 contained in the flue gas that is introduced to the carbonator:

$$\Phi = \frac{F_R}{F_{CO2}} \tag{3-8}$$

The molar flux of Ca-species that are fed to the carbonator is calculated by the chemical composition of the sorbent and the characteristic curve of the screw conveyor. The characteristic curve was empirically determined by a previous study, at cold conditions and limestone as feeding material. The following equation was derived [217]:

$$\dot{m}_S = 110.1 \frac{\rho_{sample}}{\rho_{limestone}} u_{screw} \tag{3-9}$$

Here, $\dot{m}_S$ is the total mass flow of solids transported from carbonator to calciner, ρ_{sample} is the mass density of the corresponding sample, $\rho_{limestone}$ is the mass density of the limestone and, u_{screw} is the rotating speed of the screw conveyor. This approach assumes that the mass flow of solids being transported to the calciner is equal to the mass flow of solid being fed to the carbonator. During some operation points, the L-valve was installed instead of the screw conveyor. Here, the circulating mass flow was determined by a heat balance calculation. Knowing the circulating mass flow and the corresponding mass fraction of CaO (x_{CaO}), F_R is derived as follows:

$$F_R = \dot{m}_S \frac{x_{CaO}}{M_{CaO}} \tag{3-10}$$

The CaL process heat ratio, HR_{CaL}, describes the heat intensity of the calciner. It is calculated according to Eq. 3-11.

$$HR_{CaL} = \frac{P_{th,calc}}{P_{th,carb,equiv} + P_{th,calc}} \tag{3-11}$$

Here, $P_{th,calc}$ is the thermal duty of the calciner calculated according to Eq. 3-12, and $P_{th,carb,equiv}$ is the thermal equivalent duty of the carbonator. The latter parameter describes the thermal duty of the combustion to deliver the same quality (i.e. concentration) and quantity (i.e. flow rate) of the flue gas that is fed to the carbonator.

The CaL process heat ratio is mainly influenced by the mass flow of solid circulating between carbonator and calciner and by the mass flow of fresh make-up limestone. With regard to the calciner thermal duty, lower values of solid circulation rate and make-up limestone mass flow are favorable. In addition to that, the oxyfuel combustion conditions are of concern. A low recirculation rate of the dilution agent (i.e. flue gas) and a low oxygen-to-fuel-ratio reduce the sensible heat demand of the gas flow through the calciner.

$$P_{th,CaL} = \dot{m}_{SRF} \cdot LHV_{SRF} \tag{3-12}$$

Here, $\dot{m}_{SRF}$ is the mass flow of SRF, and LHV_{SRF} is the lower heating value of SRF.

Based on the heat duty of the calciner, one can derive the specific heat demand per unit of CO_2 that is captured, Q_{spec}:

$$Q_{spec} = \frac{P_{th,calc}}{\dot{m}_{CO2,capt}} \tag{3-13}$$

Here, $\dot{m}_{CO2,capt}$ is the mass flow of CO_2 that is captured.

Similarly, the specific oxygen demand per CO_2 that is captured, $\dot{m}_{O2spec}$, is calculated as follows:

$$\dot{m}_{O2,spec} = \frac{\dot{m}_{O2}}{\dot{m}_{CO2,capt}}$$

(3-14)

Here, $\dot{m}_{O2}$ is the mass flow of O_2 that is introduced into the calciner.

3.5 Range of Process Conditions

Table 3-11 summarizes the range of relevant process parameters during the course of the experiment. The numbers are given for the overall CaL system and for each CFB sub-system respectively.

Table 3-11: Range of experimental operation conditions during pilot testing.

System	Parameter	Unit	Range
Global	Specific make-up ratio, Λ	-	0.067 - 0.323
	Specific circulation rate, Φ	-	8.88 - 27.1
	Total CO_2 capture rate, E_{tot}	%	78 - 95
Carbonator	Average riser temperature, T_{carb}	°C	640 - 700
	Superficial gas velocity, $u_{0,carb}$	m/s	2.7 - 3.1
	Molar CO_2 flux, F_{CO2}	kmol/m²s	2.5 - 3.3
	Reactor solid inventory, $W_{s,carb}$	kg/m²	257 - 576
	Equivalent thermal power, $P_{th,carb,equiv}$	kW$_{th}$	279 - 378
	CO_2 absorption efficiency, E_{carb}	%	50 - 87
Calciner	Average riser temperature, T_{calc}	°C	799 - 892
	Superficial gas velocity, $u_{0,calc}$	m/s	4.1 - 5.6
	CO_2 outlet concentration, $y_{CO2,calc,out}$	vol.%	31 - 59
	Reactor solid inventory, $W_{s,calc}$	kg/m²	133 - 458
	Thermal power, $P_{th,calc}$	kW$_{th}$	445 - 800
	Oxygen mass flow, $\dot{m}_{O2}$	kg/h	112 - 178
	Fuel mass flow, $\dot{m}_{SRF}$	kg/h	75 - 161

Based on the numbers presented above, Figure 3-7 shows an experimental summary in terms of CO_2 outlet concentrations of carbonator and calciner as a function of the corresponding average riser temperatures. The chemical equilibrium concentration of CO_2 and the corresponding reaction regimes for carbonation and calcination are indicated additionally.

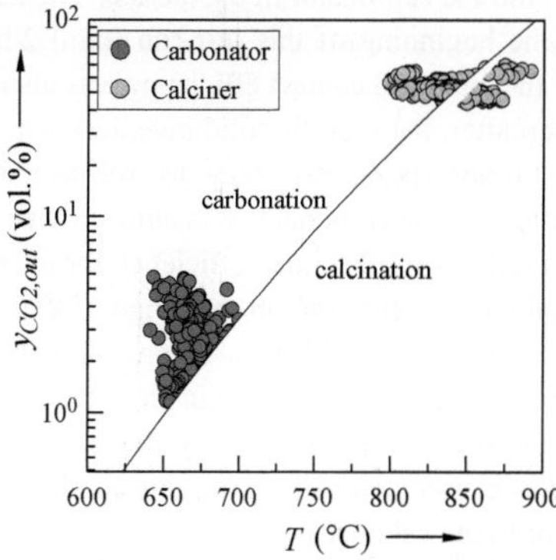

Figure 3-7: Experimental operation range of carbonator and calciner.

The average carbonator riser temperature was between 640 and 700 °C. The outlet concentration of CO_2 at the carbonator was relatively close to the corresponding chemical equilibrium line in the majority of the evaluation points. This results in CO_2 absorption efficiencies that range from 50 up to 87 %. The calciner was operated between 799 and 892 °C. During some operation points, the calciner temperature was insufficient to operate in the calcination regime. Slow calcination or re-carbonation may thus occur at calciner conditions on the carbonation side of the equilibrium curve. This was in part accepted in order to achieve a wider range of process data for the assessment of SRF combustion in the CaL calciner (see Chapter 3.10.1.5). It is important to note that each reactor performance is also influenced by the temperature profile of the riser, as the local reaction regime might differ from the one derived based on the average temperature. A detailed assessment of carbonator and calciner performance is carried out in Chapter 3.9 and 3.10.

It is desirable to achieve a highly concentrated CO_2-product stream at the outlet of the calciner to minimize the purification and upgrading demand of the GPU. However, due to practical limitations set by the 1 MW_{th} pilot plant, the CO_2 outlet concentration of the calciner was between 37 and 55 vol.% during the majority of the operation points. The dilution of the product stream is mainly caused by the presence of water vapor, nitrogen and oxygen. The fraction of water vapor is related to the combustion of SRF and is in fact easy to remove via condensation. The presence of nitrogen is associated with the required flushing of measurement devices and the fluidization of the loop seal and the cone valve. The fraction of oxygen is related to the oxygen-to-fuel, which was kept between 1.1 and 1.8 in order to ensure a nearly complete oxidation of the volatile matter of SRF (see Chapter 3.10.1.5).

Figure 3-8 shows an exemplary long-term data plot for a stable CaL operation period of 25 hours. The plotted values represent a five minutes average. Graph a) includes the average riser temperature and the specific solid inventory of carbonator and calciner, whereas Graph b) shows the superficial gas velocities of both CFBs and the molar flux of absorbed CO_2 as well as the total CO_2 capture rate. During this period, the calciner was oxy-fired by SRF I and the flue gas to be decarbonized in the carbonator was based on the combustion of natural gas. The solid mass flow between carbonator and calciner was realized by means of the L-valve.

Throughout the whole period, the carbonator was operated at a temperature of approximately 680 °C and without active reactor cooling. Approximately 80 % of the CO_2 contained in the flue gas (2.5 mol/m²s) was absorbed within the carbonator in that period. The temperature of the calciner was between 830 and 850 °C. In the beginning of this test run (until 2 h), the solid inventory in the carbonator and more precisely the gas/solid contact efficiency was insufficient to achieve the desired CO_2 absorption efficiency. Thereafter, the specific solid inventory was kept within the desired range of 400 to 500 kg/m² and the CO_2 absorption efficiency, as well as the total CO_2 capture rate, were stable. The specific solid inventory of the carbonator was almost twice as high as that of the calciner in order to account for the required gas/solid contact efficiency for a reasonable CO_2 absorption. The present conditions led to a total CO_2 capture rate in the range of 88 - 92 %, while the nominalized CO_2 absorption efficiency was between 90 and 100 %. At operational hour 17, an interruption of the SRF feed to the calciner occurred because of a blockage in the fuel feeding system. As a consequence, the calciner temperature and the superficial gas velocity decreased rapidly. Furthermore, calciner solid entrainment was reduced as consequence of the missing fluidization agent. Solid inventory was thus shifted from the carbonator to the calciner.

As discussed above, the operation of the calciner was challenging due to inhomogeneous fuel quality. The operation parameters of the pilot plant were steadily modified in order to maintain the desired process conditions for CO_2 capture. However, even with these boundary conditions, it was successfully proven that continuous CO_2 capture by means of the CaL process fired with SRF is feasible.

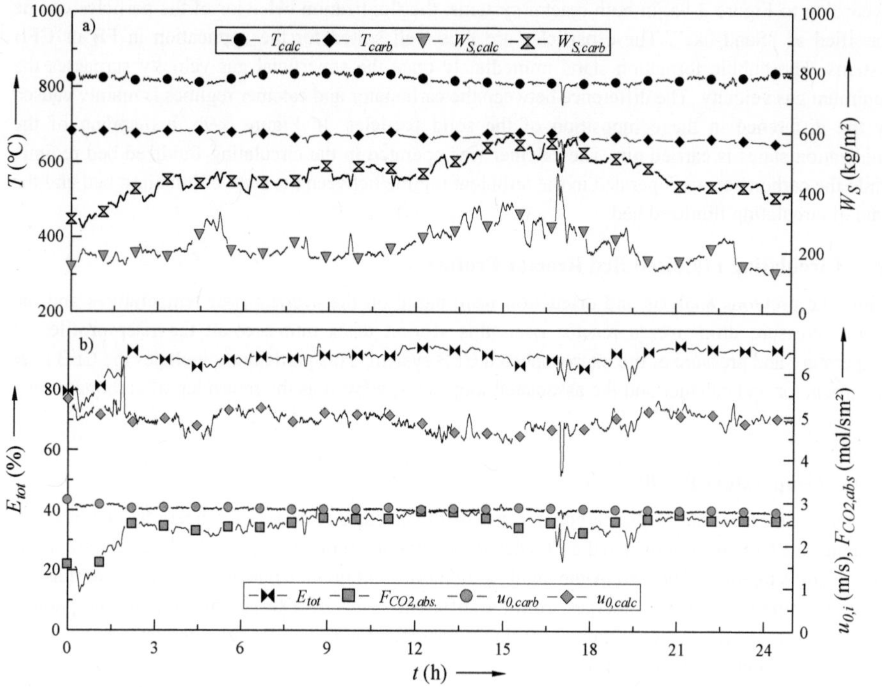

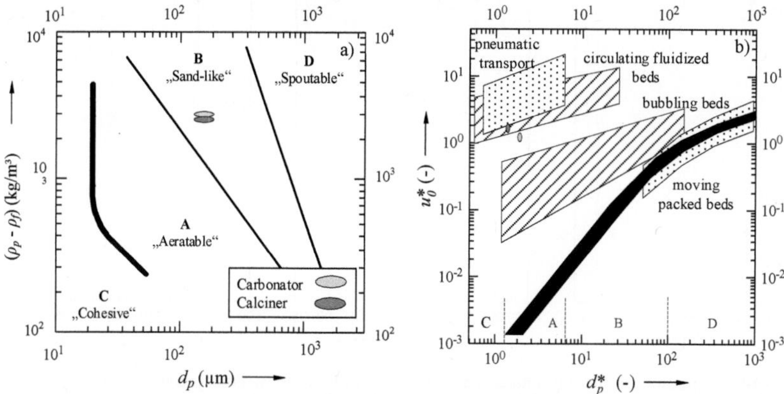

Figure 3-8: Long-term data plot. Graph a: Average reactor temperatures (T) and specific solid inventories (W_s), Graph b: Total CO_2 capture rate (E_{tot}), molar CO_2 absorption ($F_{CO2,abs}$) and superficial gas velocities (u_0).

Figure 3-9 shows the hydrodynamic characterization of the operation conditions for carbonator and calciner in terms of particle classification and fluidization state mapping (see Chapter 2.3.1 and 2.3.2). Here, the operational conditions of Table 3-11 are applied, while a $CaCO_3$ mass fraction of 20 wt.% for the carbonator and 5 wt.% for the calciner are assumed.

Figure 3-9: Hydrodynamic characterization of the experiment for carbonator and calciner. Graph a: Particle classification [83], Graph b: Fluidization states [86].

According to Figure 3-9a, in both reactor systems, the fluidization behavior of the particles can be classified as "Sand-like". These particles are thus well suited for the application in FB or CFB systems. The bubble formation starts immediately once the superficial gas velocity surpasses the minimum gas velocity. The difference between the carbonator and calciner regimes is mainly caused by the difference in the composition of the solid particles. In Figure 3-9b, a mapping of the fluidization states is carried out. The calciner was operated in the circulating fluidized bed regime, while the carbonator was operated in the turbulent regime between the state of bubbling bed and the state of circulating fluidized bed.

3.6   Circulating Fluidized Bed Reactor Profiles

While the previous analysis and discussion were based on the average riser temperatures and the overall pressure drop over a reactor riser, this section takes into account the riser profiles of temperature and pressure of the interconnected CFB system. This particularly includes the CFB riser of carbonator and calciner and the associated loop seals, as well as the return leg of carbonator and calciner.

3.6.1   Temperature Profiles

The temperature profile of the carbonator and of the calciner riser are of concern as they directly influence the CO_2 absorption and the CO_2 desorption pattern in these reactors. The temperature of the carbonator is mainly influenced by the highly exothermic carbonation reaction. In contrast to that, the calciner temperature profile emerges from the different stages of SRF combustion and the parallel occurring sorbent calcination. Figure 3-10 shows the temperature profiles for carbonator and calciner for three different process conditions.

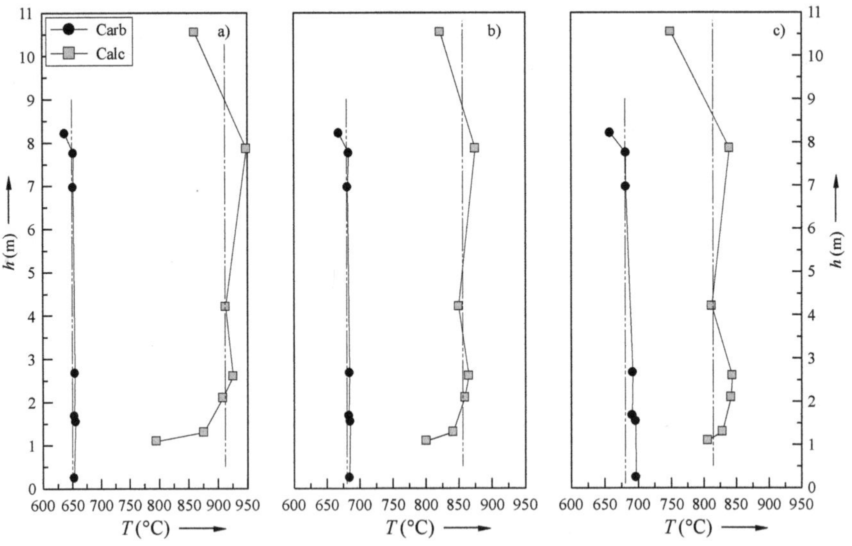

Figure 3-10: Temperature profiles of carbonator and calciner riser for different process conditions ($E_{carb,nom}$ = 62.1 %, Φ = 12.8 (a); $E_{carb,nom}$ = 96.9, Φ = 13.7 (b); E_{carb} = 89.7, Φ = 18.2 (c)).

The temperature profile of the carbonator is highly homogeneous and shows only few areas with deviations from the average riser temperature. Under the process conditions of Graph a, the measured temperature values range from 638 up to 656 °C, with the lowest value at the top of the reactor, while the bottom region shows the highest temperature. This characteristic is similar to the other cases. The temperature decrease at the top of the reactor is mainly caused by the tip of the cooling tubes which were still installed at the reactor top level. The fact that the bottom region shows a higher temperature is related to the incoming hot solids from the calciner, and to the ongoing carbonation reaction. It was already found that in the bottom region of the carbonator riser, approximately 85 % of the total CO_2 absorption occurs; consequently a large amount of heat is released in this part of the reactor [173]. Nevertheless, the carbonator holds a higher solid inventory than the calciner, which causes a more homogenous temperature profile.

In contrast to the carbonator, the temperatures of the calciner riser vary by 150 K in case of the process conditions of Graph a). The development of the profile is similar among all cases. The bottom region is relatively cold ~ 800 °C, which is mainly due to solid stream from the carbonator that is fed to the calciner at this point. Before the particles enter the riser, they are mixed with the internally circulating solids from the calciner and the SRF, thus the temperature is partly increased before this mixture is finally introduced to the riser. Here, the temperature increases as a consequence of the SRF oxidation. In the middle of the riser, temperature decreases slightly (15 - 20 K), while a higher temperature is measured at 8 m height. This decrease might be related to the ongoing endothermic sorbent calcination and the feed of cold secondary air. At the top of the calciner, the temperature decreases again as heat losses and the further heat consumption by the sorbent calcination dominate relatively to the heat release by the combustion of the remaining unconverted combustible matter.

3.6.2   Pressure Profile

Based on the pressure profiles, the solids hold up for each subsystem (e.g. riser, loop seal) could be derived. Considering the pressure differences at certain process locations, the hydrodynamic process stability and operability is assessed. In the course of this evaluation, an exemplary pressure profile of the interconnected CFB system is shown in Figure 3-11.

The values represent an average over an operation period of one hour. During that period, the calciner was fired by SRF I, and the pilot plant was operated stably both, from a hydrodynamic and CO_2 absorption point of view. The solid flux from carbonator to calciner was controlled by means of a screw conveyor. The specific solid inventories were approximately 300 and 570 kg/m² for calciner and carbonator, respectively. The characteristic pressure profile of a CFB reactor is pronounced for carbonator and calciner. A distinct dense region in the bottom part of each reactor (a) and the above lean region can be distinguished for each riser. Both reactors were operated with internal solid circulation, in order to decouple the particle entrainment rate from the global solid circulation rate. The pressure in each loop seal (c) exceeds the pressure through the reactor riser at that height, which ensures the pressure sealing for the gas flow.

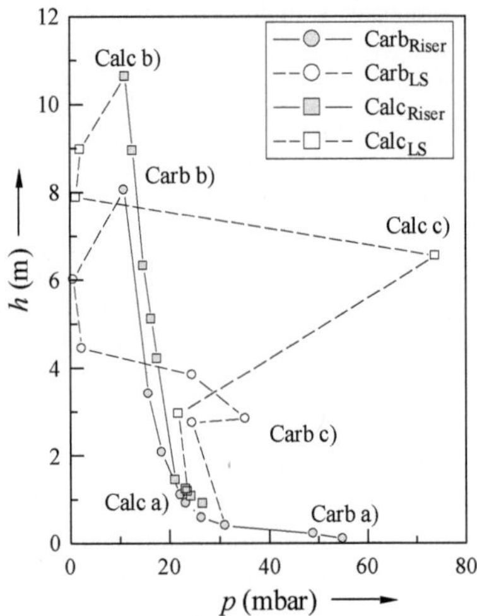

Figure 3-11: Pressure profile of the interconnected CFB system, CaL process configuration with screw conveyor (a: Nozzle grid, b: Reactor outlet, c: Loop seal).

3.7 Sorbent Properties

During the experimental test runs, numerous solid samples were discharged from the CaL pilot plant at relevant process locations. After the experiment, relevant solid samples that were taken during stable evaluation points have been identified. These samples were analyzed with various principles in order to gain additional information on CaL process performance (see Chapter 3.4.1). In this chapter, the chemical composition of the sorbent, its particle size distribution as well as its activity and mechanical stability are presented and discussed.

3.7.1 Chemical Composition

The extent of relevant species in the circulating solid streams is of concern for process evaluation, as the operational conditions of carbonator and calciner are simultaneously affected and reflected by the various species contained in the solid phase. Furthermore, the extent of a secondary utilization of purged material is limited by the presence of solid species, such as heavy metals, sulfur and chlorine. Some substances, such as ash or $CaSO_4$, remain stable during the operation conditions of the CaL process, whereas the content of $CaCO_3$ varies according to the performance of carbonator and calciner. Figure 3-12 depicts the mass fraction of relevant solid streams for a representative operation period while SRF I was fired in the calciner.

This particular CaL operation point is characterized by a total CO_2 capture rate of 93.9 %, whereas the calcination efficiency was 90.5 % which results in a remaining $CaCO_3$ content at the calciner outlet of 2.2 wt.%. During the CO_2 absorption in the carbonator, the $CaCO_3$ mass fraction increases again up to 21.02 wt.%. Interestingly, the bottom ashes discharged from carbonator and calciner show a relatively high mass fraction of $CaCO_3$. In case of the carbonator, this phenomena is likely related to the fact of accumulation and further carbonation of coarse Ca-species in the bottom region of the reactor. Given the higher time for carbonation, carbonate content increases up to 34.6 wt.%. For the calciner, the lower temperature in the bottom region (see Chapter 3.6.1) and the high CO_2

concentration in the fluidization gas are the major cause for the increase in the carbonate content up to 26.1 wt.%.

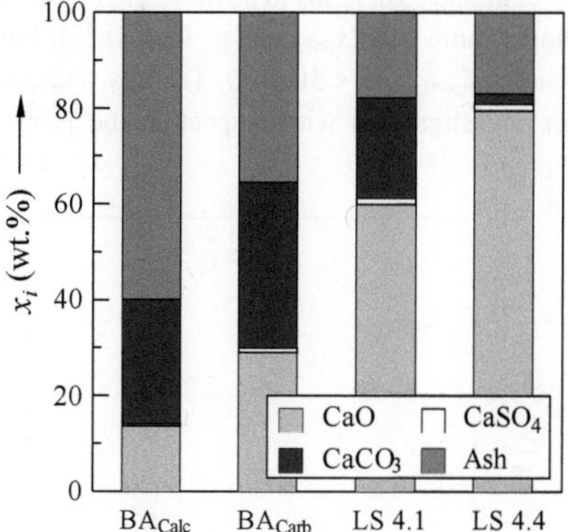

Figure 3-12: Sorbent composition of CaL material streams. Fuel: SRF I, E_{carb} = 84.9 %, E_{tot} = 93.9 %, E_{calc}= 90.5 % (BA: Bottom ash, LS 4.1: Carbonator outlet, LS 4.4: Carbonator inlet).

The presence of ash in the different solid streams varies significantly. In the circulating sorbent stream (i.e. sampling points LS 4.1 and LS 4.4), the mass fraction of ash is at a similar level ($x_{Ash,LS\,4.1}$ = 16.9 wt.%, $x_{Ash,LS\,4.4}$ = 17.8 wt.%), whereas it varies considerably between the two bottom ash streams. The bottom ash of the calciner contains the largest share of ashes ($x_{BA,Calc}$: 59.9 wt.%) as a consequence of the accumulation of the coarse fraction in the bottom of the calciner. Additionally, there is a relatively large ash fraction present in the carbonator bottom ash ($x_{BA,Carb}$: 35.5 wt.%), since there is also the tendency of coarse ash accumulation. This phenomenon is sustained by the fact that the carbonator superficial gas velocity is approximately 3 m/s, while the calciner superficial gas velocity increases by the factor of 1.67. The particle size distribution of coarse ash fraction in carbonator and calciner bottom is further addressed in Chapter 3.7.2.

Only a minor share of SRF ash is considered as typical fly ash which is continuously entrained throughout the calciner cyclone separator. The main share of SRF ash accumulates in the solid looping cycle either at the bottom of the CFB reactors (coarse ashes) or in the circulating sorbent stream (finer ashes). The quantity of ash in the circulating sorbent mainly depends on the size distribution of the fuel ash, on the limestone make-up feeding rate and on the fluidization conditions in the solid looping system. The latter parameter has been kept relatively constant, so that the ash concentration is minimized, once the limestone make-up feed is maximized.

Figure 3-13 shows the mass fraction of ash in the circulating sorbent dependent on the make-up feed related to the SRF mass flow for SRF I and II, respectively. The values given for the ash mass fraction represent an average of the numbers related to the samples extracted at the sampling points LS 4.1 and LS 4.4.

Throughout the experimental investigations, the mass fraction of ash varies between 17.2 and 10.5 wt.% for SRF I and between 16.8 and 12.4 for SRF II. For both types of SRF, the observed trend is in accordance with the intrinsic fundamentals of the CaL process. Once more limestone make-up is introduced to the process per quantity of SRF, more ashes are discharged from the solid looping system. Among the two types of SRF, the fuel analysis in terms of ash mass fraction and the particle size specification are of concern for the assessment of the ash accumulation. Accordingly, SRF I shows a lower ash fraction ($x_{Ash,SRF\ I}$ = 9.70 wt.%) and a lower maximum particle size ($d_{p,max,SRF\ I}$ < 30 mm), thus the contained ash tends to be more likely present in the circulating sorbent. Controversially, SRF II contains more ash ($x_{Ash,SRF\ II}$ = 15.4 wt.%), but the entrainment of ash is limited by the large particle size ($d_{p,max,SRF\ II}$ < 50 mm). To fully understand the fate of ashes in the solid looping system, further investigations with respect to the properties of the fuel ashes are required.

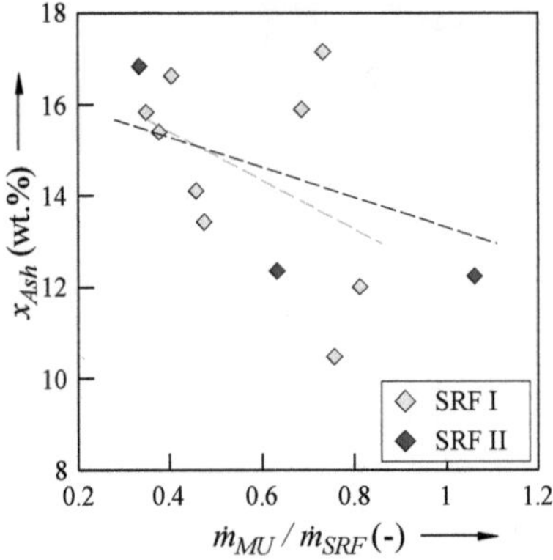

Figure 3-13: Mass fraction of ash in the circulating sorbent (x_{ash}) dependent on the SRF-specific make-up feed ($\dot{m}_{MU}/\dot{m}_{SRF}$).

$CaSO_4$ represents another disadvantageous sorbent constitute. In addition to its inert nature, the sorbent deactivation is forced by the presence of $CaSO_4$ (see Chapter 2.5.4.2). The sorbent is exposed to sulfur in the carbonator (flue gas SO_x) and in the calciner (fuel sulfur). Figure 3-14 shows the mass fraction of $CaSO_4$ in the circulating sorbent as a function of the SRF-specific make-up feed for SRF I and SRF II. Furthermore, the figure distinguishes between low sulfur (synthetic, natural gas derived) and high sulfur (LEP-derived) flue gas in the carbonator, respectively.

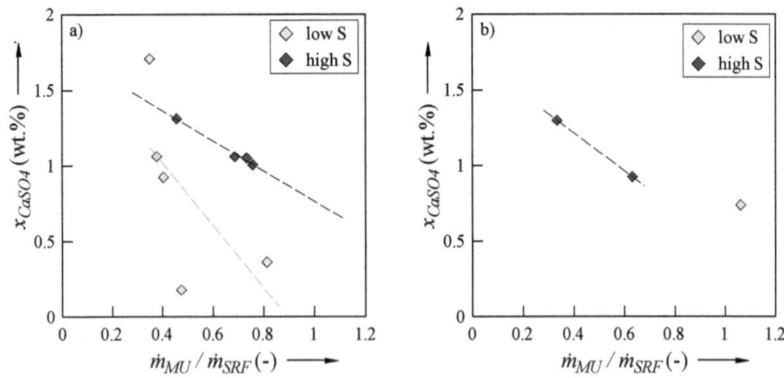

Figure 3-14: Mass fraction of calcium sulfate in the circulating sorbent (x_{CaSO4}) dependent on the SRF-specific make-up feed ($\dot{m}_{MU}/\dot{m}_{SRF}$) for operation points firing SRF I (a) and SRF II (b).

The sulfur mass fraction in the circulating sorbent varies between 0.18 wt.% (low-sulfur FG, SRF I, $\dot{m}_{MU}/\dot{m}_{SRF} = 0.47$) and 1.71 wt.% (low-sulfur FG, SRF I, $\dot{m}_{MU}/\dot{m}_{SRF} = 0.35$). Similarly to the behavior of the ash mass fraction in the circulating sorbent stream, a higher SRF-specific make-up rate favors lower $CaSO_4$ loads. Overall, the utilization of SRF I allows for a lower sulfur accumulation thanks to the lower specific sulfur concentration of the fuel in comparison to SRF II ($x_{S,spec,SRF\,I} = 0.08$ g/MJ$_{th}$, $x_{S,spec,SRF\,II} = 0.19$ g/MJ$_{th}$). Nevertheless, it needs to be noted that more data points, especially for operation points firing SRF II, are required to fully verify this statement. The effect of the SO_x quantity in the flue gas to be decarbonized in the carbonator on the $CaSO_4$ enrichment is evident in Figure 3-14a. Here, the trend line referring to low-sulfur flue gas data points is clearly at a lower level compared to the high-sulfur data points.

3.7.2   Particle Size Distribution

With knowledge of the PSD of the various material streams, conclusions can be drawn about the abrasion resistance of the limestone or the accumulation of coarse components within solid looping system. Figure 3-15 shows the PSD of selected solid samples for a stable operation period firing SRF I in the calciner.

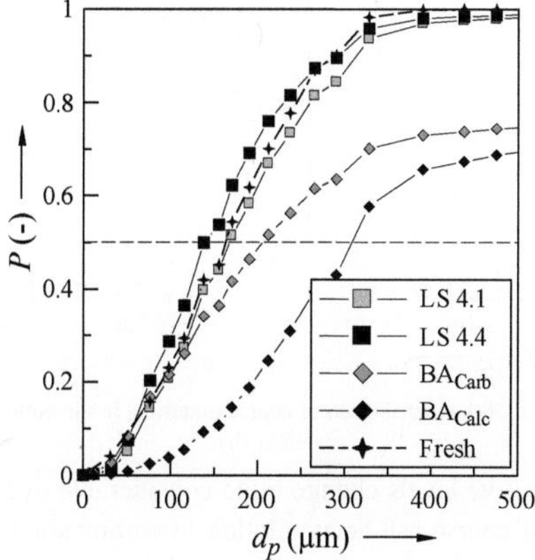

Figure 3-15: Exemplary PSD of material streams while firing SRF I in the calciner (LS: Loop Seal, BA: Bottom ash, Fresh: Limestone make-up).

The mean particle diameter of the circulating sorbent stream is in the range of approximately 140 to 160 µm, which represents a reduction of approx. 15 - 35 µm relatively to the mean particle diameter of the fresh make-up limestone. The sample extracted from the outlet of the calciner at LS 4.4 shows a smaller particle size in comparison to the samples extracted from the outlet of the carbonator at LS 4.1. This phenomena might be related to several circumstances. As for instance to the point of the make-up feed or to the lower resistant of the calcined sorbent to mechanically induced attrition as a consequence of the calcination. Furthermore, the cyclone particle separation efficiency further influence the PSD of the associated solid stream.

Contrary to the PSD of the circulating sorbent stream, the PSD of the bottom ashes tend to increase during operation. The mean particle diameter of the carbonator bottom ash increases to 205 µm, while the mean particle diameter of the calciner bottom ash increases to 305 µm. This phenomena is clearly assigned to the coarse ash fractions of SRF. As already noted, these materials are to a large extend hardly fluidizable and accumulate at the bottom of the calciner and carbonator. The carryover of these materials from calciner to carbonator is due to differences in the related superficial gas velocities. The carbonator was typically operated with a lower superficial gas velocity than the calciner. Thus, coarse particles might be fluidizable under conditions of the calciner, but not under conditions of the carbonator.

Waste-derived fuels contain relevant components of coarse, non-combustible fractions, such as metal, glass and mineral matter. Depending on the size specification of the fuel, these inert constituents tend to negatively influence the plant operation due to several reasons. The change of the fluidization behavior in the CFB reactors and in the loop seals due to accumulation of coarse components serve as an example. Furthermore, there is the risk of a blockage of the ash extraction system by wires or large metal pieces. However, the latter issue might be of less importance in an appropriately designed large-scale CaL system. The distribution of the coarse ash fractions over an operation period of 80 hours during the first CaL test campaign is shown in Figure 3-16.

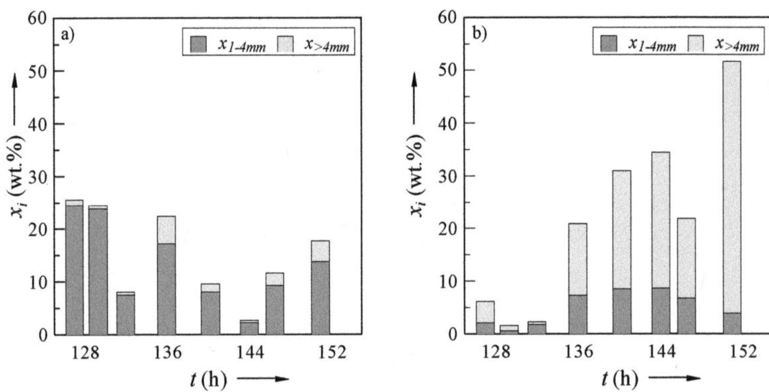

Figure 3-16: Time development of the distribution of coarse materials in the bottom ashes of carbonator (a) and calciner (b).

Even though the total quantitative levels change quite considerable over time, there is a clear trend that proves the occurrence of coarse ash accumulation in carbonator and calciner. While particles with a size of 1 - 4 mm occupy an average of 13.3 wt.% of the carbonator bottom ash, this number is significantly lower among particles larger than 4 mm ($x_{> 4mm,av}$ = 1.95 wt.%). For the calciner, this relation shows a contrary pattern. Accordingly, large ash factions are associated with 16.3 wt.% of total bottom ash, while particles in the range of 1 - 4 mm are associated with 4.95 % of total bottom ash. The fluctuations over time are mainly related to the heterogeneous nature of SRF.

3.7.3 Sorbent Stability

The mechanical stability of the sorbent is relevant for the techno-economic process performance of the CaL process (see Chapter 2.5.4.5). The minimum required make-up feed is affected by the extent of the sorbent losses caused by elutriation of fine particles that cannot be kept in the solid looping system by the cyclone separators. This in fact influences the operational expenses, due to the costs related to the supply of fresh limestone. The attrition coefficient is relevant to predict the lifetime of the sorbent.

Figure 3-17 shows the development of the sorbent abrasion coefficient and the increment sorbent abrasion coefficient of solid samples extracted from LS 4.1 (carbonator outlet) and LS 4.4 (carbonator inlet) in the course of the first CaL test campaign.

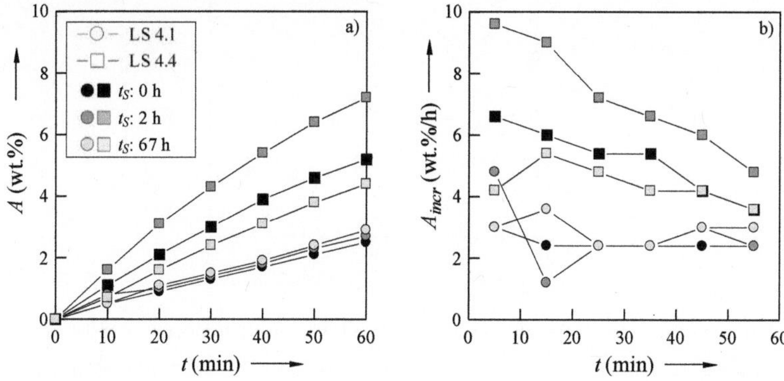

Figure 3-17: Development of the sorbent abrasion coefficient (a) and increment sorbent abrasion coefficient (b) for sorbent samples taken at different times (t_S) at the carbonator inlet (LS 4.4) and the carbonator outlet (LS 4.1).

The total cumulative abrasion coefficient at the end of attrition test varies between 2.5 and 2.9 wt.% among the samples taken at the outlet of the carbonator. This parameter ranges from 4.4 to 7.2 wt.% for the sorbent at the outlet of the calciner. There is a clear distinction of the abrasion coefficient between solid samples taken from LS 4.1 and LS 4.4. Accordingly, the partially carbonated sorbent at the outlet of the carbonator is less prone to the mechanical abrasion phenomenon. Furthermore, the abrasion resistance of the calcined sorbent is more associated with the fluctuating operation conditions of the calciner, as the total cumulative abrasion coefficient is more scattered in contrast to the partially carbonated sorbent samples.

Figure 3-17b shows the increment sorbent abrasion coefficient at each time of sample mass weighing during the abrasion tests. Neglecting the effects of the initial abrasion in the first 30 minutes, it can be stated that the increment abrasion coefficient of the partly carbonated sorbent remains constant, while this parameter decreases along with increasing test duration for the calcined sorbent.

As already noted in Chapter 2.5.4.5, the presence of the carbonate product layer increases the mechanical resistance of the sorbent. This fact is graphically proven in Figure 3-18 for sorbent samples taken at the outlet of the calciner (LS 4.4) and of the carbonator (LS 4.1), respectively.

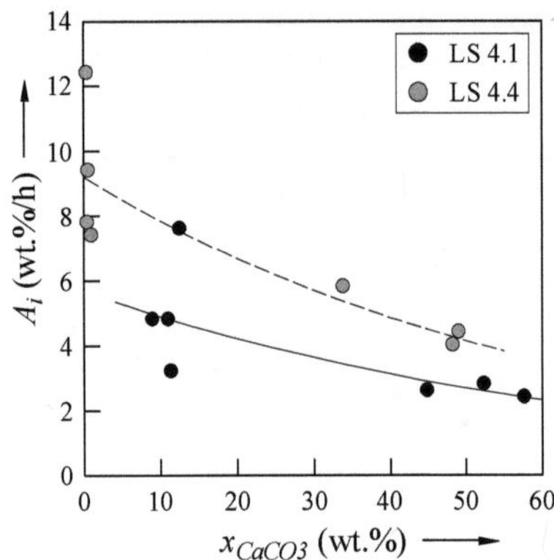

Figure 3-18: Attrition coefficient (A_i) as a function of the $CaCO_3$ mass fraction (x_{CaCO_3}) of the sorbent.

For both data sets, a higher $CaCO_3$ mass fraction clearly reduces the material losses assigned to the mechanical abrasion phenomena. It needs to be noted that some of the solid samples were taken during operational periods, when the calciner temperature was insufficient to achieve a reasonable high calcination efficiency. Typically, there is almost no carbonate product layer available on the sorbent particles at the outlet of the calciner. Nevertheless, these circumstances further prove the positive effect of the carbonate product layer on the abrasion resistance.

Knowing the abrasion coefficient of the sorbent, it is possible to estimate the reduction of the mean particle diameter of the circulating sorbent. In qualitative terms, a higher abrasion coefficient is associated with a forced reduction of the mean particle diameter. Figure 3-19 shows the relative reduction of the mean particle diameter as a function of the abrasion coefficient. The mean particle diameter is derived from the pre-sieved sorbent fraction.

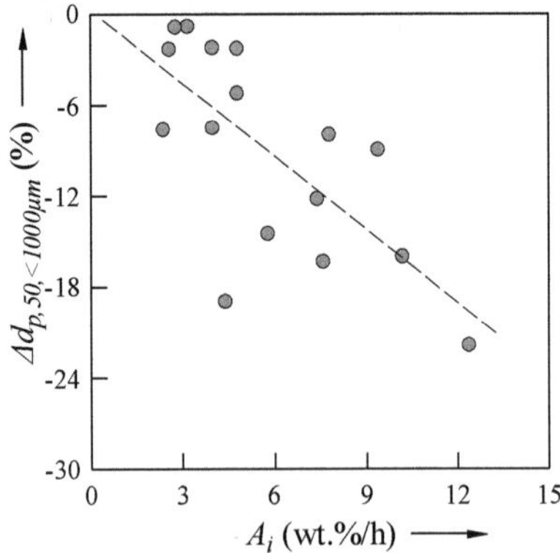

Figure 3-19: Relative reduction of the mean particle diameter of the pre-sieved sorbent fraction ($\Delta d_{p,50,<1000\mu m}$) dependent on the abrasion efficient (A_i).

The reduction of the mean particle diameter ranges from 0.84 to 21.9 %, referring to the initial mean particle diameter of 179 µm. There is a reasonable correlation in terms of mean particle diameter reduction and sorbent abrasion coefficient. Nevertheless, it needs to be noted that in addition to the particle attrition, other effects such as ash accumulation or the gas velocity in the cyclone separators influence the mean particle diameter of the circulating sorbent.

3.7.4   Sorbent Activity

The sorbent activity describes the ability of the circulating sorbent to absorb CO_2 by means of the carbonation reaction in the carbonator. More precisely, the molar conversion of the sorbent at the transition point between the "fast" (limited by the reaction kinetics) and the "slow" (limited by the diffusion resistance of the product layer) carbonation reaction regime is assigned to the activity of the sorbent (see Figure 2-15) [152].

In order to assess the deactivation characteristics of the sorbent, the sorbent activity of selected samples is drawn as a function of the prevailing CaL process conditions at the time of sample extraction. Figure 3-20 emphasizes the dependency of the specific make-up ratio for two distinguished calciner temperature ranges. It needs to be noted that a lower calciner temperature regime might be inherently associated with an insufficient sorbent calcination (see Chapter 3.10). The deactivation patterns shown by the data set referring to $T_{calc,high}$ are thus more evident to express the expected conditions in a large-scale CaL unit.

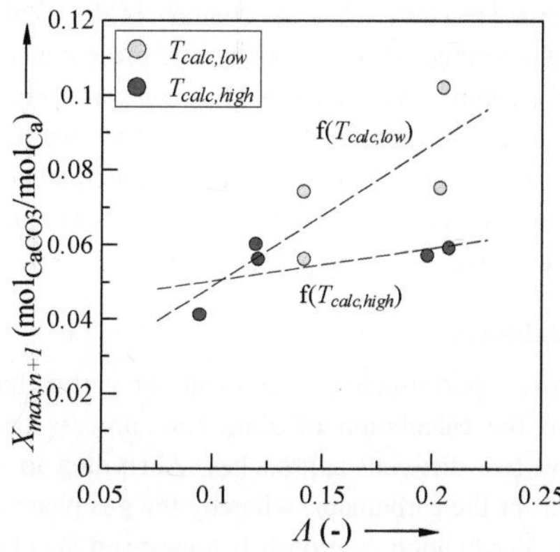

Figure 3-20: Sorbent activitiy ($X_{max,n+1}$) dependent on the specific make-up ratio (Λ) for different average calciner temperatures ($T_{calc,low}$ < 836 °C, $T_{calc,high}$ > 836 °C).

It can be seen that the activity of the sorbent is favored by low calciner temperatures as well as by high specific make-up rates. The activity of the analyzed solid samples ranges from 0.041 to 0.102 mol_{CaCO3}/mol_{Ca}. In the relevant calciner temperature range for the CaL process, an activity level of 0.041 to 0.060 mol_{CaCO3}/mol_{Ca} prevails. The average number of complete carbonation-calcination cycles is reduced once more fresh make-up limestone is fed to the process. This is proven by Figure 3-20 as the activity increases along with higher specific make-up feeding rates. This dependency is more pronounced in case of the low calciner temperature range.

In addition to the number of carbonation-calcination cycles and the conditions during sorbent calcination, the sulfation level of the sorbent is a major cause for sorbent deactivation. This phenomenon is associated with the blockage of the particle pores by $CaSO_4$ (see Chapter 2.5.4.2). Figure 3-21 shows the sorbent activity as a function of the $CaSO_4$ mass fraction of the sorbent.

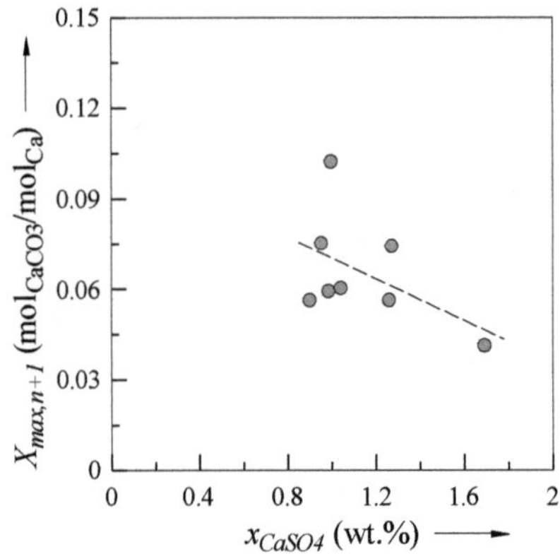

Figure 3-21: Sorbent activity ($X_{max,n+1}$) dependent on the calcium sulfate mass fraction (x_{CaSO4}).

The sorbent activity ranges from 0.041 to 0.102, while the associated $CaSO_4$ mass fraction spans from 0.90 to 1.69 wt.%. In accordance with the theoretical fundamentals, the lowest sorbent activity is related to the highest $CaSO_4$ mass fraction, while the contrary is almost also the case. At this point, it is required to raise the need for further investigations, as the present analysis is potentially prone to uncertainties. For example, the sample mass used in TGA analysis represents only a minor share of the total circulating sorbent stream. Furthermore, the type of fuel, the residence time of particles in the calciner, as well as the prevailing gas atmosphere, are parameters that influence the sorbent deactivation. Finally, the total number of analyzed solid samples is rather low, thus the total solid inventory might not be fully acquired.

3.8 Closure of Carbon Balances

As a first step of the CaL process performance assessment, the carbon balances need to be verified, as this metric is relevant to the calculation of almost all process performance indicators. The verification is carried out by two different approaches. According to Verification Approach I, a carbon balance is established for the carbonator, whereby the gas phase and the solid phase carbon flows are taken into account. Verification Approach II focuses on the closure of the carbon balance for the calciner.

3.8.1 Verification Approach I

By this method, the differences of carbon flows at the inlet and at the outlet of the carbonator are compared for gas and solid phase respectively. The CO_2 absorbed from the gas phase in the carbonator needs to be equal to the CO_2 in the form $CaCO_3$ newly bound in the solid phase:

$$E_{carb}F_{CO2,carb,in} = F_R(X_{carb} - X_{calc}) \qquad (3\text{-}15)$$

The left hand term of the above equation is derived from online gas analysis and gas flow measurement. The carbon built up in the solid phase is determined by the molar flow of Ca-species fed to the carbonator, F_R, and the carbonate content at the carbonator inlet, X_{calc}, and outlet, X_{carb}. The circulating mass flow is calculated by the characteristic curve of the screw conveyor or by a heat balance calculation in case of the L-valve. The molar flow of the Ca-species is derived from the sorbent composition and the total mass flow of the circulating sorbent. The right hand term is based on solid sampling and subsequent solid analysis. Thus, the number of available data point is limited by the numbers of solid samples that were evaluated. Figure 3-22 shows the closure of the carbon balance for the carbonator during stable operation points.

The amount of CO_2 absorbed in the carbonator ranges from 1.56 to 2.35 mol/m²s. For a large part of the evaluation points, there is a reasonable agreement (+/- 30 %) between the gas phase and solid phase carbon balance. Inaccuracies are partially related to the fact that the solid samples taken during the experiment represent only a minor share of the total circulating mass flow and the total solid inventory of the system. In this particular case, the evaluation was even more challenging due to the inhomogeneous fuel quality, which led to fluctuations in the calciner operation conditions and consequentially in the composition of the solid phase and the derived solid phase carbon balance. In this regard, a reasonably long pilot plant operation without modification of operation conditions would be ideally for a representative sorbent evaluation. The solid phase balance slightly tends towards higher numbers for the specific amount of CO_2 being absorbed in contrast to the value indicated by the gas phase measurement. Likely, this trend is associated with the evaluation of the solid phase. The calculation of the solid circulation rate is a rather difficult task in interconnected CFB systems.

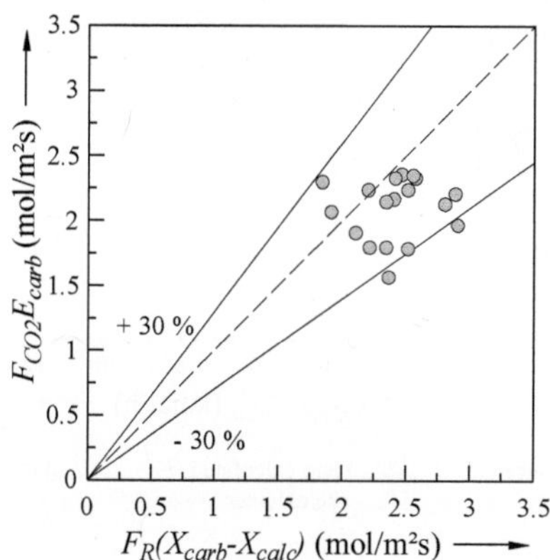

Figure 3-22: Closure of carbonator carbon balance for CaL evaluation points.

3.8.2   Verification Approach II

This method assesses the flow rate of CO_2 at the outlet of the calciner, e.g. downstream of the extraction point of the recirculation gas. This metric represents the flow rate of CO_2 that is captured by the CaL process. It could be either directly derived from the measurement devices at the pilot plant, or indirectly calculated from the sum of all the CO_2 present in the outlet of the calciner as a function of the carbon-containing feed streams to the CaL calciner, e.g. CO_2 absorbed in the carbonator, limestone make-up, SRF, flushing agent. Ideally, both approaches show the same value:

$$F_{CO2,capt,meas} = F_{CO2,capt,calcu} \tag{3-16}$$

$$F_{CO2,calc,out} - F_{CO2,calc,reci} = F_{CO2,abs} + F_{CO2,SRF} + F_0 + F_{CO2,flush} \tag{3-17}$$

The measured molar flux of captured CO_2, $F_{CO2,capt,meas}$, is compared to the sum of all molar fluxes of CO_2 which will be captured according to the intrinsic CaL process characteristics, $F_{CO2,capt,calcu}$. These molar flows are the CO_2 that is absorbed in the carbonator, $F_{CO2,abs}$, the CO_2 released by the combustion of SRF, $F_{CO2,SRF}$, and the CO_2 associated with the calcination of the make-up limestone, F_0 as well as the CO_2 flux used for flushing or fluidization purposes at the calciner, $F_{CO2,flush}$. For the calculation of the right hand term in the above equation, complete fuel oxidation and limestone calcination in the calciner are assumed. The comparison of the molar CO_2 fluxes that are captured is shown in Figure 3-23.

The carbon balance of the majority of operation points is well verified by +/- 20 %. Furthermore, all points are reasonably equal divided along the line, which implies that there is no systematic inaccuracy with regard to the evaluation methodology. Major cause for inaccuracies here is the heterogeneous nature of SRF. As the elementary composition of the fuel fluctuates over time, it directly affects the calciner off-gas quality and quantity. However, the evaluation is based only on one exemplary SRF composition data set, thus the variations in SRF quality are not fully taken into account.

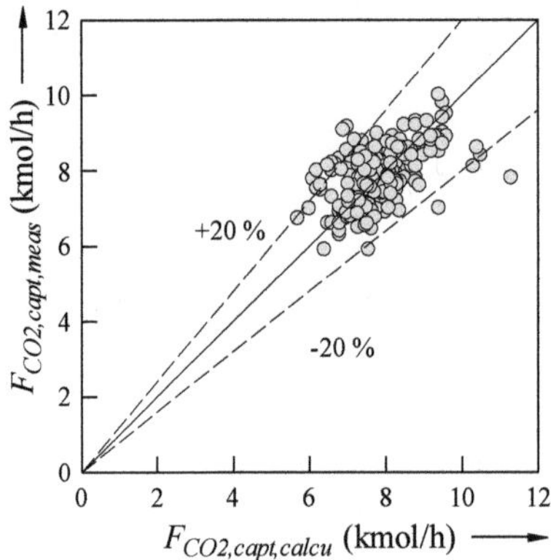

Figure 3-23: Comparison of measured ($F_{CO2,capt,meas}$) and calculated ($F_{CO2,capt,calcu}$) molar fluxes of CO_2 at the outlet of the calciner.

3.9 Carbonator Operation

The carbonator performance (i.e. the CO_2 absorption efficiency) is a key metric for the CaL process as it significantly alters the total CO_2 capture rate. Several parameters influence the carbonator performance, either directly (e.g. specific solid circulation rate, specific solid inventory, average carbonator riser temperature or specific make-up rate) or indirectly (e.g. type of supplementary fuel, extent of SO_x in the flue gas to be decarbonized). This evaluation covers an assessment of the influence of the average carbonator riser temperature, the specific make-up and sorbent circulation rate. Furthermore the effect of the specific carbonator solid inventory and the carbonator active space time on the CO_2 absorption efficiency in the carbonator are addressed.

3.9.1 Effect of the Carbonator Temperature

The average carbonator riser temperature influences the CO_2 absorption in two contrary ways. While a higher temperature allows for faster carbonation reaction kinetics, the chemical equilibrium in the CaO-$CaCO_3$ systems shifts towards CaO and the theoretically maximum feasible CO_2 absorption efficiency is reduced accordingly.

Figure 3-24 shows the carbonator CO_2 absorption efficiency dependent on the average carbonator riser temperature. Additionally, the theoretical maximum CO_2 absorption efficiency at chemical equilibrium conditions is marked as a solid line based on an inlet CO_2 concentration of 9.5 vol.%. The operation points are distinguished according to the type of flue gas that was fed to the carbonator.

The majority of the test points is in the temperature range of 660 to 680 °C yielding CO_2 absorption efficiencies of 60 to 80 %. In this temperature range, the limiting effect of the chemical equilibrium is noticeable through the reduced absorption efficiency along with a higher average carbonator temperature. Operation points that show a higher CO_2 absorption than the theoretical maximum value are either based on a higher CO_2 inlet concentration than in the baseline case, or additional absorption of CO_2 occurs in the upper part of the reactor, where the temperature is usually lower than the average value (see Chapter 3.6.1). The latter phenomenon was already identified by means of in-bed measurement of the CO_2 concentration in the carbonator riser [173]. Once the carbonator temperature is lower than 660 °C, the CO_2 absorption efficiency dropped below 70 % for the majority of test points in that range. It can therefore be concluded that a carbonator temperature of approximately 660 °C is favorable in terms of CO_2 absorption efficiency. Furthermore, it needs to be taken into account that a higher carbonator operation temperature inherently lowers the heat demand of the calciner, which results in a lower specific heat demand for the CO_2 capture process (see Chapter 3.11.3). On the basis of the available data from pilot plant operation, no significant difference in terms of CO_2 absorption efficiency patterns between the three types of flue gas is noticeable.

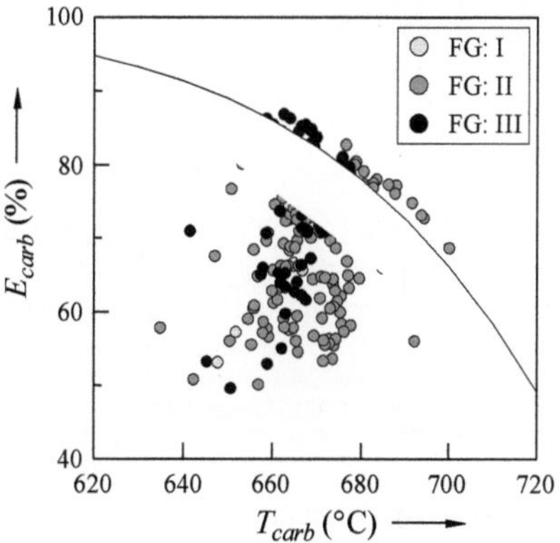

Figure 3-24: Carbonator CO_2 absorption efficiency (E_{carb}) dependent on the average carbonator temperature (T_{carb}) for different types of flue gas (FG I: synthetic, FG II: NG, FG III: LEP).

3.9.2 Effect of the Specific Make-up Rate

The constant feed of fresh limestone is necessary in order to maintain the desired CO_2 carrying capacity of the circulating sorbent. Additionally, the accumulation of inert species in the solid loop is avoided by the constant replacement of bed material. It is desirable to keep the make-up feeding rate as low as possible, since the heat of reaction associated with the first calcination of the limestone cannot be recovered. Furthermore, costs for the supply of make-up needs to be accounted for in the cost balance of the CaL process. The make-up feeding rate is typically expressed as the molar ratio of Ca-species contained and the CO_2 present in the flue gas fed to the carbonator (Λ, see Eq. 3-7). In previous investigations of the CaL process at the same pilot plant using lignite and hard coal as supplementary fuels in the calciner, this parameter was kept between 0.04 and 0.25 [174]. For full-scale CaL applications, the specific make-up rate is assumed to range from 0.03 to 0.38 (see Table 2-5).

Figure 3-25 shows the nominalized CO_2 absorption efficiency as a function of the specific make-up rate of the experimental tests with SRF. Here, it is distinguished between data points with a low and a high specific circulation rate (Φ_{low}, Φ_{high}), respectively.

For the considered data points, the specific make-up rate was varied between 0.116 and 0.323. Nominalized CO_2 absorption efficiency varies between 63.1 and 100 %. When more make-up is fed to the process, the average maximum sorbent activity increases. Accordingly, more CO_2 is absorbable in the fast reaction stage of sorbent carbonation, which is beneficial with regard to the CO_2 absorption efficiency. For all herein considered data points, a higher specific make-up feed results in an improved CO_2 absorption performance. This relation is further assessed by means of thermogravimetric analysis in the course of Chapter 3.6. Among the majority of data points, a specific sorbent circulation rate below ~ 12 is not sufficient to achieve nominalized CO_2 absorption efficiencies above 90 %, even at relatively high specific make-up rates.

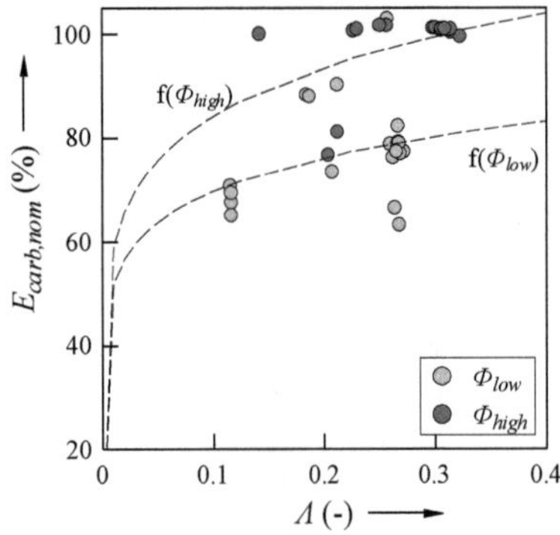

Figure 3-25: Nominalized CO_2 absorption efficiency ($E_{carb,nom}$) dependent on the specific make-up rate (Λ), $W_{s,carb}$ = 257-364 kg/m², Φ_{low} < 12, Φ_{high} > 15.

3.9.3   Effect of the Specific Sorbent Circulation Rate

The specific sorbent circulation rate (Φ) takes into account the Ca-species contained in the solid stream that is introduced to the carbonator (F_R). Similarly to the specific molar make-up ratio, this parameter is expressed in relation to the CO_2 contained in the flue gas that is fed to the carbonator (F_{CO2}, see Eq. 3-8). At the 1 MW$_{th}$ pilot plant, this parameter can be adjusted by the rotating speed of the screw conveyor or in case of an L-valve, by the appropriate setting of the fluidization gases. The specific sorbent circulation rate has a significant influence on the heat balance of the CaL system, as this stream needs to be heated in the calciner and subsequently cooled in the carbonator ($\Delta T_{(calc-carb)}$ ~ 250 K). For full-scale CaL process applications, the specific sorbent circulation rate is assumed to range from 3.0 up to 12.6 (see Table 2-5).

Figure 3-26 shows the nominalized CO_2 absorption efficiency dependent on the specific sorbent circulation rate. It needs to be noted that at higher specific sorbent circulation rates, the nominalized carbonator CO_2 absorption efficiency for the high specific make-up rate case would always be higher than for the medium specific make-up rate case contrary to the trend lines showing the mathematical fit of the data points.

Among the considered data points, the specific sorbent circulation rate was varied between 9.56 and 25.4. The associated nominalized CO_2 absorption efficiency ranges from 52.0 to 100 % (i.e. the chemical equilibrium limits any further CO_2 absorption). The positive effect of the specific sorbent circulation rate is most clearly pronounced in the medium specific make-up rate case (i.e. Λ_{med}). Contrary, in the high specific make-up rate case, the sorbent shows a higher maximum average activity. Thus, for a given circulation rate and a given gas/solid contact efficiency, more CO_2 is absorbed in the kinetically controlled fast reaction regime. On the contrary, in case of a less active sorbent (i.e. in the medium specific make-up rate case), the circulation rate needs to be increased to keep the carbonator inventory the same as active as in the high specific make-up rate case.

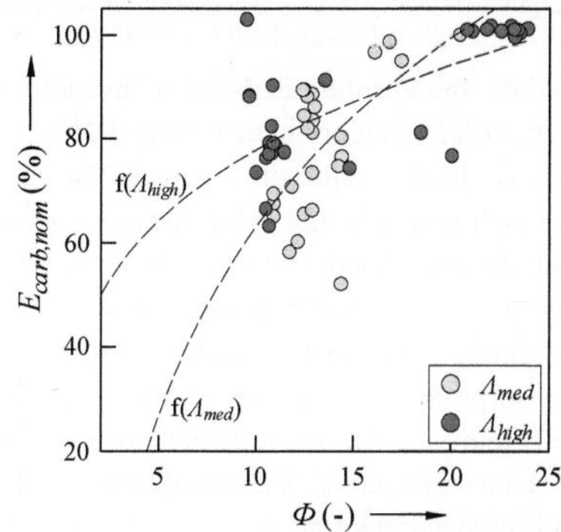

Figure 3-26: Nominalized CO_2 absorption efficiency ($E_{carb,nom}$) dependent on the specific sorbent circulation rate (Φ), $W_{s,carb}$ = 257-364 kg/m², Λ_{med}= 0.100 - 0.175, Λ_{high} > 0.175.

3.9.4 Effect of the Specific Carbonator Inventory

The specific carbonator inventory expresses the solid holdup mass per unit of carbonator riser cross-section area. The higher this number, the longer the average residence time of the particles is per carbonation-calcination cycle and the higher is the gas/solid contact efficiency. Furthermore, this value influences the pressure drop of the flue gas that passes through the carbonator, which in fact alters the auxiliary power load of the CaL unit. Thus, the appropriate solid hold up in the carbonator needs to be chosen based comprehensive approach.

Figure 3-27 shows the nominalized CO_2 absorption efficiency of the carbonator dependent on the specific carbonator solid inventory for two different specific make-up rates that range from 0.100 - 0.175 for graph a) and from 0.175 - 0.323 for graph b). Furthermore, within each graph, three different parameter ranges are distinguished in terms of the specific sorbent circulation rate (Φ_{low} < 12, Φ_{med} 12 - 15, Φ_{high} > 15).

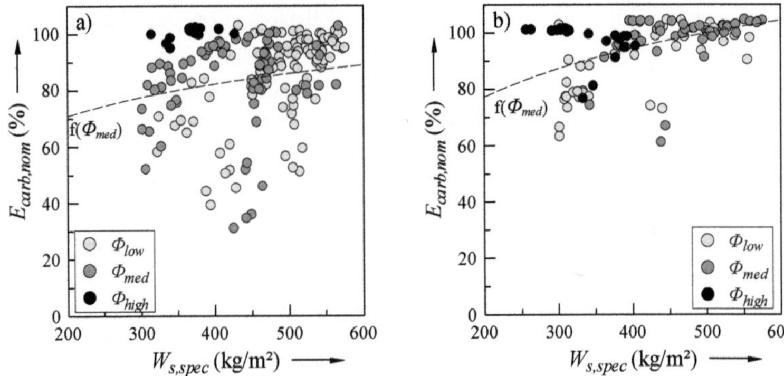

Figure 3-27: Nominalized CO_2 absorption efficiency ($E_{carb,nom}$) dependent on the specific carbonator inventory ($W_{s,spec}$), $\Lambda = 0.100 - 0.175$ (a), $\Lambda = 0.175 - 0.323$ (b).

Among the considered data points, the absolute carbonator inventory was varied between 72 and 162 kg, which results in a specific solid inventory in the range of 257 - 576 kg/m². The positive effect of the carbonator solid hold up is clearly visible in Figure 3-27a. Here, for the low and medium specific sorbent circulation rate data sets, a higher solid inventory leads to a higher CO_2 absorption efficiency. In case of the highest specific sorbent circulation rate, the influence is almost insignificant, as the supply of calcined Ca-species exceeds the molar extend of CO_2 in the flue gas by at least 15 times. It could be stated, that more than approximately 550 kg/m² of solid hold up is required to allow for a nominalized CO_2 absorption efficiency above 80 %. For the data points referring to the highest specific make-up ratio (Figure 3-27b), the influence of the specific inventory is diminished as the circulating sorbent tends to be more active. A nominalized CO_2 absorption efficiency is already feasible once the solid hold up is above 350 kg/m².

3.9.5 Effect of the Carbonator Active Space Time

In the previous assessment, the carbonator performance was evaluated by means of single parameter dependencies, which means that the influence of one single parameter on another was assessed. Due to the rather complex and multi-layered nature of the CaL process and its investigations at 1 MW_{th} scale, it is hardly possible to maintain all other operation parameters as constants while changing one single parameter. The observed trends during the analysis of a single dependency are most likely also governed by secondary dependencies. A more comprehensive evaluation methodology based on the carbonator active space time, $\tau_{carb,active}$, has thus been postulated by Charitos et al. [152].

The proposed methodology is based on simple reactor and particle reaction models applied to the solid and gas phase balance around the carbonator [152, 227]. According to Eq. 3-18, the amount of CO_2 removed from the gas phase needs to be equal to the amount of CO_2 absorbed by the active solid phase:

$$E_{carb}F_{CO2,carb,in} = n_{Ca,active}\left(\frac{dX}{dt}\right) \tag{3-18}$$

The latter term could be described by the active carbonator inventory, $n_{Ca,active}$, and an average reaction rate term (dX/dt). The average reaction rate is derived from the assumption of a plug flow reactor according to Eq. 3-19.

$$\frac{dX}{dt} = k_S \varphi X_{carb}\left(y_{CO2,carb,av} - y_{CO2,carb,eq}\right) \tag{3-19}$$

Here, k_S is the limestone specific reaction rate, and φ is the gas-solid contact factor for turbulent fluidized beds. The apparent rate constant ($k_S\varphi$) is seen as a fitting parameter and is calculated by means of an iterative comparison between solid and gas phase carbon balances. In the present analysis, a value of 0.315 1/s is assumed. X_{carb} is the molar carbonation degree of the sorbent at the outlet of the carbonator. The conditions of the gas phase are accounted for by the differences between the average CO_2 carbonator concentration, $y_{CO2,carb,av}$ and the CO_2 concentration at chemical equilibrium, $y_{CO2,carb,eq}$.

The active carbonator inventory, $n_{Ca,active}$ is calculated using the active fraction, f_{active}, of the total Ca-species in the carbonator:

$$f_{active} = 1 - e^{\frac{-t^*}{n_{Ca}/F_{Ca}}} \tag{3-20}$$

Here, t^* represents the time required to increase the molar carbonate content from X_{calc} to X_{carb} along with the reaction rate term from Eq. 3-19:

$$t^* = \frac{X_{ave} - X_{calc}}{k_S\varphi X_{ave}\left(y_{CO2,carb,av} - y_{CO2,carb,eq}\right)} \tag{3-21}$$

The carbonator active space time, $\tau_{active,carb}$, is then derived from the combination of Eq. 3-19 to Eq. 3-21. This term describes the CO_2 absorption efficiency of the carbonator dependent on (i) the carbonator solid inventory, (ii) the solid circulation rate, (iii) the concentration of CO_2 in the gas phase, (iv) the average carbonator temperature, (v) limestone specific reaction parameters and (vi) the prevailing fluidization conditions:

$$\tau_{active,carb} = \frac{n_{Ca}}{F_{CO2,carb,in}} f_{active} X_{ave} \tag{3-22}$$

Figure 3-28 shows the nominalized CO_2 absorption efficiency of the carbonator as a function of the carbonator active space time. In addition to the experimental values, the model line is plotted.

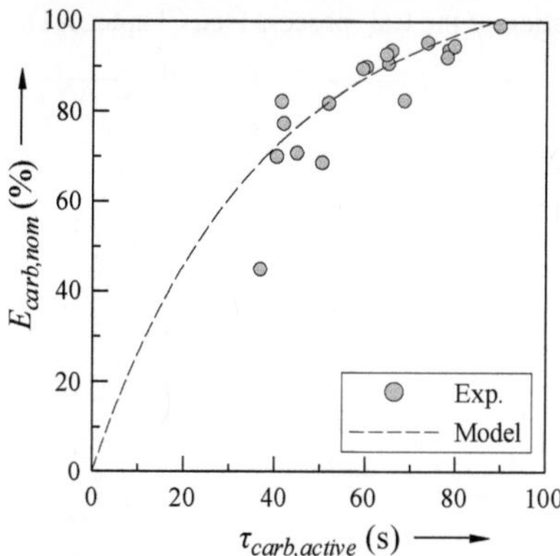

Figure 3-28: Normalized CO_2 absorption efficiency as a function of the carbonator active space time ($T_{carb} = 670$ °C, $y_{CO2,in} = 9.5$ vol.%, $k_S\varphi = 0.315$ 1/s).

There is a satisfying agreement between the model line and the experimental data despite the aforementioned difficulties with regard to solid phase analysis. The observed trend is in agreement with other experimental CaL studies carried out in different pilot plants and under varying process conditions [174, 176, 228, 229]. According to the applied evaluation methodology, a carbonator active space time of approximately 60 s is required to achieve a nominalized CO_2 absorption efficiency above 80 %.

3.10 Calciner Operation

The regeneration of the partly carbonated sorbent is carried out in the oxyfuel-fired calciner. The calcination of carbonated sorbent occurs simultaneously to the oxyfuel combustion of SRF. Therefore, the evaluation of the calciner operation is twofold. First, the formation of major gaseous pollutants associated with the combustion of SRF is investigated. Second, the calciner performance is discussed in terms of sorbent regeneration.

3.10.1 Evaluation of Gaseous Pollutant Formation

As first part of the calciner evaluation, the combustion of SRF in CFB reactor is assessed in terms of major gaseous pollutants emissions, such as CO, NO, SO_2 and HCl. Here, the experimental framework was extended in order to include the combustion of SRF in a stand-alone CFB reactor in addition to the combustion in the calciner of the CaL process. This particularly implies the utilization of silicate sand as bed material in the stand-alone CFB reactor case, in contrast to limestone in the case of the CaL calciner. By this approach, a basis for comparison of the gaseous pollutants was established. Furthermore, in addition to the combustion in a real oxyfuel environment (e.g. technically pure oxygen and recirculated flue gas), air or oxygen-enriched air was considered as the oxidation agent in all cases. Four different environments are thus available for the combustion assessment of each type of SRF.

3.10.1.1 Experimental Configuration

The experimental configuration is schematically shown in Figure 3-29. Schematic a) shows the pilot plant setup during Test Period I (TP I) for the combustion of SRF in a standalone CFB reactor. Schematic b) shows the pilot plant setup during Test Period II (TP II), which is similar to the configuration for the investigation of the CaL process (see Chapter 3.1).

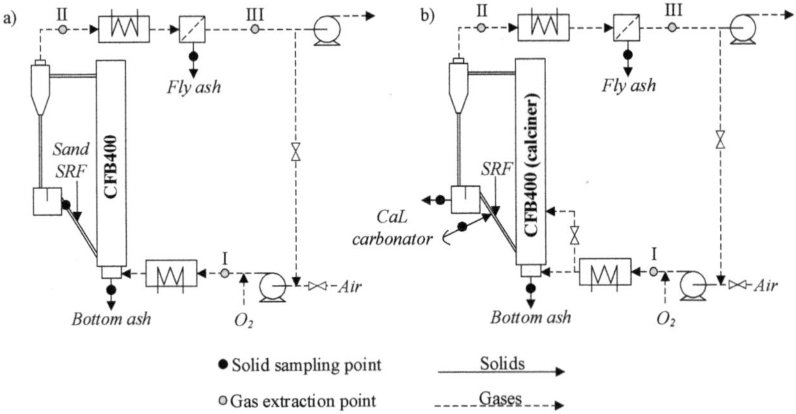

Figure 3-29: Experimental setup for the investigation of SRF combustion in a CFB combustion system (a) and in a CaL calciner (b).

During TP I, silicate sand and SRF were fed to the reactor riser via the return leg together with the internally circulating particles. As no reactor cooling system is available at the CFB400, the thermal duty was limited in that test period. Furthermore, no secondary air supply was installed during TP I. Apart from these modifications, there are no considerable differences in the experimental configuration between TP I and TP II. The applied measurement devices were already introduced in Chapter 3.2.

3.10.1.2 Materials Applied

During the combustion tests in TP I, the CFB 400 was operated with silicate sand as bed material. Table 3-12 summarizes the chemical composition of the silicate sand. The diameter of the sand particles varies between 90 µm and 355 µm with a mean particle diameter of 200 µm. The PSD and the chemical composition of the limestone and the technical properties of the two types of SRF can be found elsewhere (Chapter 3.3, Table 3-7).

Table 3-12: Chemical composition of silicate sand.

Component	Chemical formula	Mass fraction (wt.%)
Silicon dioxide	SiO_2	99.5
Iron (III) oxide	Fe_2O_3	0.04
Aluminum oxide	Al_2O_3	0.25

3.10.1.3 Evaluation Methodology

Figure 3-30 graphically shows the evaluation methodology that is applied for the evaluation of the combustion tests. Here, the evaluation methodology for oxygen-enriched air atmosphere (Schematic a) and for oxyfuel atmosphere (Schematic b) are distinguished from one another.

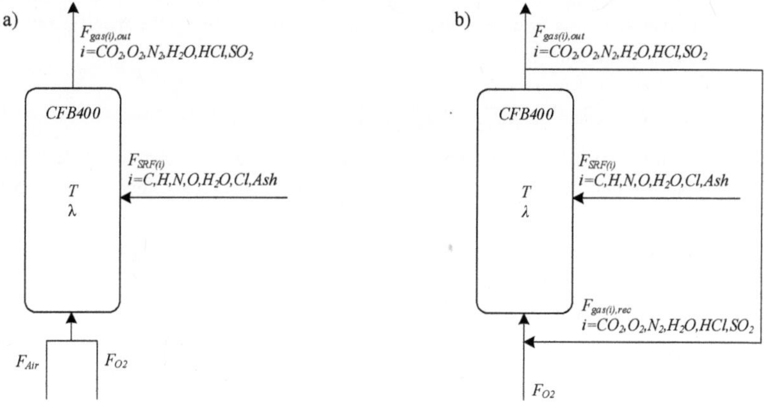

Figure 3-30: Evaluation methodology for the combustion tests in air or oxygen-enriched air atmosphere (a) and oxyfuel atmosphere (b).

In case of oxygen-enriched air atmosphere, the volumetric gas flow at the inlet represents the sum of the technically pure oxygen measured by an orifice plate, and the flow of ambient air supplied by the ID-fan measured by a venture flow measurement. Based on the gravimetrically measured mass flow of SRF and the given elementary composition, it is possible to derive the oxygen-to-fuel ratio, λ, as one crucial metric for the evaluation of combustion processes:

$$\lambda = \frac{\dot{m}_{O2,act}}{\dot{m}_{O2,min}} \qquad (3\text{-}23)$$

Here, $\dot{m}_{O2,act}$ is the actual mass flow of oxygen supplied to the reactor and, $\dot{m}_{O2,min}$ is the minimal mass flow of oxygen required for full oxidation of the organic matter (i.e. C, H, S) in SRF as calculated by Eq. 3-24.

$$\dot{m}_{O2,min} = (2.67x_C + 8x_{H2} + x_S - x_{O2}) \cdot \dot{m}_{SRF} \qquad (3\text{-}24)$$

Here, x_i is the mass fraction of the corresponding fuel species, i, of SRF. In the above equation, the oxygen present in SRF needs to be accounted for with a negative sign. The stoichiometric oxygen demand for each combustible fuel species is based on the quotient of the molar masses of oxygen and of the corresponding combustible species, multiplied by the pre-factor of the corresponding reaction stoichiometry, assuming complete oxidation.

The quantity of the flue gas and the relevant flue gas species (i.e. O_2, CO_2, N_2, H_2O, NO, SO_2, HCl) is derived from the pressure drop in a venture flow measurement and by the gas composition as measured by online gas analysis. Typically, the stack emission limits are discussed in the unit of mass pollutant per volume of flue gas at standardized conditions (i.e. mg/Nm^3). Thereby, the flue gas composition needs to be nominalized to a given volumetric oxygen concentration in order to avoid the effects of flue gas dilution as a consequence of different oxygen-to-fuel-ratios.

Under oxygen-enriched air and oxyfuel combustion conditions, the oxygen concentration at the inlet of the furnace varies as well. Thus, the dilution of the flue gas is dependent on the oxygen-to-fuel ratio and on the oxygen concentration at the inlet of the furnace. In order to avoid any misinterpretation, here, the gaseous emission pollutants are given in reference to the thermal duty of the furnace. Based on the mass flow of each species, it is possible to derive the specific emissions, e_i, of the corresponding pollutants (i.e. CO, NO, SO_2, HCl) according to Eq. 3-25:

$$e_i = \frac{\dot{m}_i}{P_{th,LHV}} = \frac{\dot{m}_i}{LHV_{SRF} * \dot{m}_{SRF}} \qquad (3\text{-}25)$$

Hereby, $\dot{m}_i$ is the mass flow of the considered pollutant species, i, downstream the cyclone separator, and $P_{th,LHV}$ is the thermal duty of the CFB furnace based on the mass flow of SRF, $\dot{m}_{SRF}$, and corresponding LHV_{SRF}.

For the oxidation products of Cl and S, it is possible to derive the corresponding retention rate of the furnace, R_i according to Eq. 3-26. Here, it is assumed that the gaseous emission species are released exclusively in the form of SO_2, as in the case of sulfur, or HCl, as in the case of chlorine. Given the fact that part of the SO_2 further oxidizes to SO_3 which is not measured, the retention rate of sulfur tends to be overestimated. It was found that the total SO_3 conversion ratio is below 1 % in case of oxyfuel CFB combustion of coal [230].

$$R_i = 1 - \frac{F_{gas,out,i}}{F_{i,SRF}} = 1 - \frac{F_{gas,out,i}}{x_{i,SRF} * \dot{m}_{SRF} * \frac{1}{M_i}} \qquad (3\text{-}26)$$

In case of oxyfuel atmosphere, the above outlined methodology needs to be partly modified. The quantities of gaseous species require to be reduced by the amount of flue gas that is recirculated to the inlet of the furnace. Doing so, each species is considered to remain inert in the recirculation pathway. Consequently, the oxygen-to-fuel ratio at the inlet of the furnace needs to be corrected by the recirculated flow of oxygen.

3.10.1.4 Range of Experimental Conditions

Table 3-13 provides an overview of the operational conditions relevant for the evaluation of SRF combustion in the 1 MW_{th} CFB pilot plant. During all test periods, the oxyfuel combustion of SRF was the main scientific purpose. However, in order to provide a better understanding of the emission formation characteristics, air and oxygen-enriched air were also considered as oxidation agents.

Due to the heat consumption of the endothermic limestone calcination and the cooling of the reactor by the incoming, relatively cold sorbent from the carbonator, the thermal duty of the furnace was significantly higher during TP II in comparison to TP I. Furthermore, the supply of the oxidation agent was exclusively carried out through the nozzle grid of the riser during TP I.

Table 3-13: Range of experimental conditions during TP I and TP II.

Parameter	Unit	TP I	TP II
Bed material, -	-	Sand	Ca-species
Temperature CFB400, T	°C	780 - 880	750 - 900
Thermal power, P_{th}	kW_{th}	290 - 440	450 - 820
Superficial gas velocity, u_0	m/s	3.7 - 5.0	4.5 - 6.0
Specific solid inventory, $W_{s,spec.}$	kg/m²	375 - 850	150 - 400
O_2 inlet concentration, $y_{O2,in}$	vol.%$_{dry}$	21 - 33	40 - 55
CO_2 inlet concentration, $y_{CO2,in}$	vol.%$_{dry}$	0 - 35	0 - 30
Secondary air ratio, SAR	%	-	30 - 45
Stoichiometric O_2-ratio, λ	-	1.1 - 1.9	1.1 - 1.9
Experimental duration, t	h	20	230

3.10.1.5 Influence of Process Conditions on Gaseous Pollutant Emissions

In the course of this chapter, the emissions of major gaseous pollutants (i.e. CO, NO, SO_2, HCl) are evaluated for each type of SRF in each combustion environment. The main purpose is the characterization of the gaseous emissions under different process conditions (e.g. temperature, oxygen-to-fuel-ratio). Ultimately, it is desirable to conclude under what conditions the formation of certain species is minimized. With regard to the CaL process, this is of concern as the process conditions can be optimized in order to allow for a reasonable CO_2 capture performance while minimizing the upgrading requirement of the flue gas stream in the GPU.

3.10.1.5.1 Carbon Monoxide

Figure 3-31 shows the nominalized specific CO emissions as a function of the oxygen-to-fuel ratio for all combustion types of SRF and combustion environments. The top line (Graph a and b) of this figure shows the specific CO emissions during the combustion in a sand-based CFB and the bottom line (Graph c and d) shows the specific CO emissions evolving from SRF combustion in the environment of the CaL calciner. The overall level of specific CO emissions was relatively high throughout all experimental investigations, which indicates a poor fuel burnout. This is justified by several explanations.

The first reason is associated with the feeding point of SRF to the riser (see Chapter 3.1). The fuel is introduced to the return leg of the CFB400. In the return leg, the fuel is mixed together with the internally circulating flow of hot solids. Thus, the release of volatile matter already starts in this part of the furnace. Thereafter, the mixture of gaseous and solid fuel compounds is introduced to the riser slightly above the nozzle grid at one point of the circumference. At this point it is likely that part of the already released volatile matter flows upwards in the wall zone of CFB reactor without getting in sufficient contact with the oxidation agent. Thereby, a certain split stream of CO passes through the riser and contributes significantly to the overall CO emission level.

The second reason is related to the feeding system for SRF. As introduced in Chapter 3.1, the mass flow rate of SRF bases on the velocity of a belt conveyor and on the height of SRF on the belt. Due to low mass density of fluffy SRF, the feeling level in the hopper of the SRF feeder considerably influences the degree of SRF compression on the moving belt. Thus, in case of relatively empty hopper, the SRF mass flow rate tends to be lower, while in case of relatively full hopper, the SRF mass flow tends to be higher. This issue was already addressed during operation by the selection of a narrow filling level, but it cannot be completely avoided due to the characteristics of the SRF feeder.

The third reason is associated with the fuel itself. Due its fluctuating quality and composition, the carbon distribution between volatile matter and fixed carbon varies considerably. This theoretically requires a steady modification of the CFB400 combustion settings during operation.

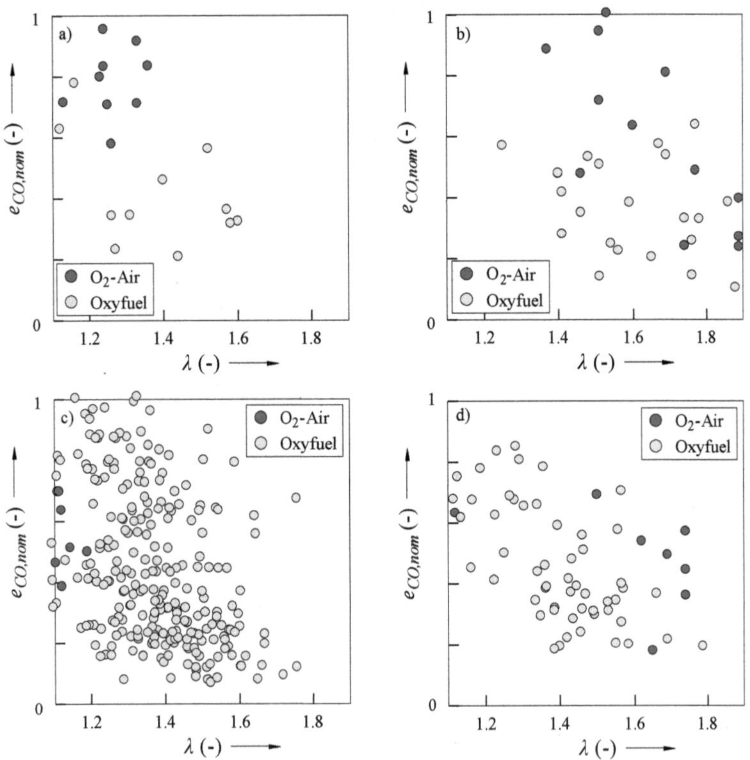

Figure 3-31: Nominalized specific CO emission ($e_{CO,nom}$) dependent on the oxygen-to-fuel ratio (a: SRF I, TP I; b: SRF II, TP I; c: SRF I, TP II; d: SRF II, TP II).

Throughout all combustion environments, the oxygen-to-fuel ratio significantly influences the CO emission level. An oxygen-to-fuel ratio larger than approximately 1.5 is required to achieve sufficiently low CO emission levels. The positive effect of the secondary air injection is particularly visible in comparison of the graphs b) and d). Here, the combustion environment of the CaL calciner, clearly favors lower CO emissions in contrast to the combustion in the CFB400 without secondary air injection. Given the fact, that the experimental investigations were mainly attributed to the CO_2 capture performance of the whole unit, operation conditions were altered accordingly.

3.10.1.5.2 Nitrogen Oxide

The main process parameter that influence the formation of NO is the oxygen-to-fuel-ratio (λ). Figure 3-32 shows the specific emissions of NO (e_{NO}) in the course of TP I (top line) and of TP II (bottom line) as a function of the oxygen-to-fuel-ratio.

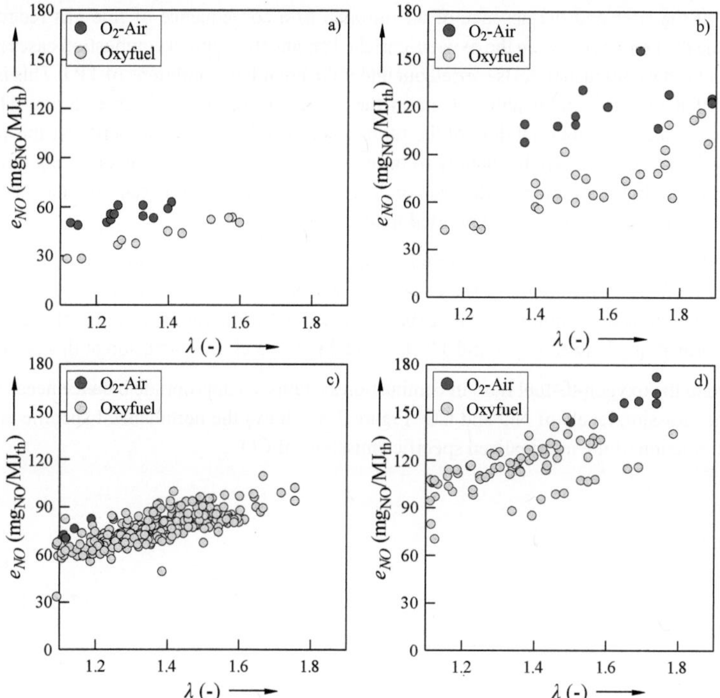

Figure 3-32: Specific NO emission (e_{NO}) as a function of the oxygen-to-fuel ratio (a: SRF I, TP I; b: SRF II, TP II; c: SRF I, TP II; d: SRF II, TP II).

As expected, a higher oxygen-to-fuel ratio favors the emissions of NO throughout all combustion environments and types of fuel. During TP I, the specific NO emission range from 28 mg$_{NO}$/MJ$_{th,LHV}$ (SRF I, oxyfuel) up to 155 mg$_{NO}$/MJ$_{th,LHV}$ (SRF II, oxygen-enriched air). The corresponding numbers for TP II range from 55 mg$_{NO}$/MJ$_{th,LHV}$ (SRF I, oxyfuel) and 172 mg$_{NO}$/MJ$_{th,LHV}$ (SRF II, oxygen-enriched air). The volumetric concentrations of NO are higher in an oxyfuel combustion environment compared to the case of oxygen-enriched air, due to a lower pollutant dilution in the oxygen-enriched air case. Exemplary numbers for TP I range from 91 ppm$_v$ (SRF I, oxygen-enriched air) up to 561 ppm$_v$ (SRF II, oxyfuel). In the course of TP II, the NO concentration in the flue gas range from 174 ppm$_v$ (SRF I, oxygen-enriched air) to 607 ppm$_v$ (SRF II, oxyfuel).

Throughout all combustion tests, the combustion of SRF II tends towards higher specific NO emissions. This phenomenon is partially justified by the differences in the elementary composition of both types of SRF. More precisely, the nitrogen mass concentration of SRF II (0.97 wt.%$_{db}$ = 0.62 g$_N$/MJ$_{th,LHV}$) is twice as high as that of SRF I (0.44 wt.%$_{db}$ = 0.21 g$_N$/MJ$_{th,LHV}$). This further proves the evidence of the fuel-N formation pathway with regard to the total NO emission during FB combustion of SRF, even under conditions of a CaL calciner. However, the differences between both types of fuels are more pronounced in case of TP I.

Furthermore, the specific NO emissions tend to be lower in the oxyfuel case for both types of SRF as well as during both test periods. Major reasons for that are the reduction of recirculated NO in the lower part of the riser and the absence of air nitrogen as a consequence of flue gas recirculation. However, the differences between the oxygen-enriched air and the oxyfuel combustion cases are more pronounced for the combustion tests carried out under the boundary conditions of TP I. This is related to the fact that the fraction of ambient air in the oxidation agent was higher during TP I as a consequence of the lower thermal duty of the furnace compared to TP II. Furthermore, this might be justified by the fact that the whole quantity of the oxidation agent is supplied exclusively through the nozzle grid at the bottom region of the reactor. The oxidation of the nitrogen volatile species (e.g. NH$_3$ and HCN) towards NO is thus supported in that region.

When assessing the effect of the bed material on the emission formation propensity, there is a slight tendency of higher NO formation in the case of a Ca-species based solid inventory. At this point, further investigations are required, as the experimental framework varies in terms of secondary air ratio and thermal duty between TP I and TP II, which hinders a clear conclusion at this point.

With regard to the oxygen-to-fuel ratio in combustion systems, an appropriate balance needs to found between the emission levels of NO and CO. Figure 3-33 shows the nominalized specific emissions of NO as a function of the nominalized specific emissions of CO.

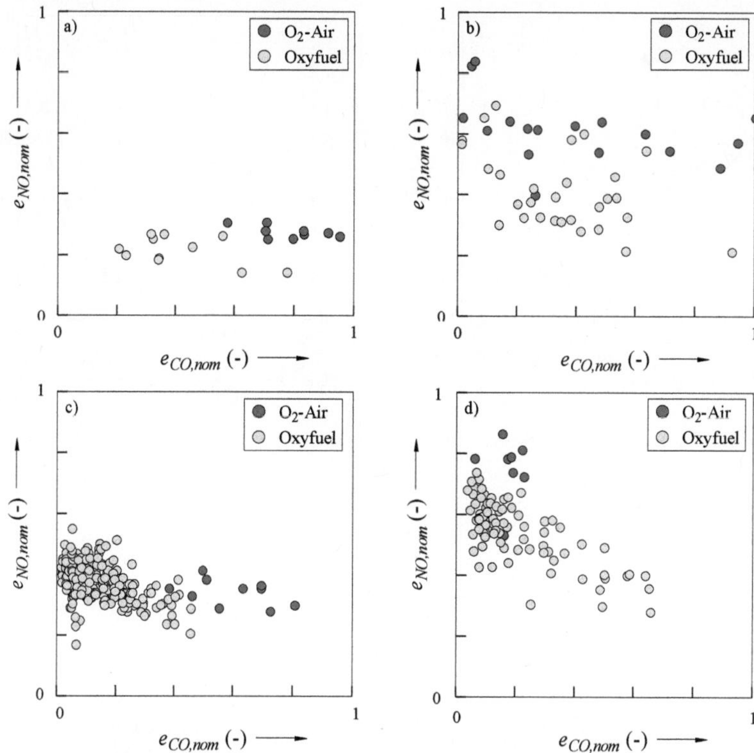

Figure 3-33: Nominalized specific NO emission ($e_{NO,nom}$) as a function of the nominalized specific CO emissions ($e_{CO,nom}$) during TP I (a: SRF I, b: SRF II) and during TP II (c: SRF I, d: SRF II).

It is known that the combustion condition related to both emission species influence each other in an opposite way. With regard to the oxygen-to-fuel ratio, higher values are here favorable in order to keep the CO emission level low, while in parallel NO emission level increases. Contrary, a low combustion temperature favors moderate NO emission level, while the fraction of CO in the flue gas increases. These fundamental connection is visible throughout the SRF combustion in a stand-alone CFB (top line of Figure 3-33) and throughout the SRF combustion in the framework of the CaL calciner (bottom line of Figure 3-33). Furthermore, the relation between NO and CO emission level is more pronounced in case of SRF II (Graphs b and d) as the content of fuel-nitrogen is higher (see Chapter 3.10.1.5.1).

3.10.1.5.3 Sulphur Dioxide

Figure 3-34 shows the specific SO_2 emissions (e_{SO2}) and the derived sulfur retention rates (R_S) as a function of the furnace temperature for the combustion of SRF I (top line) and SRF II (bottom line) during TP I.

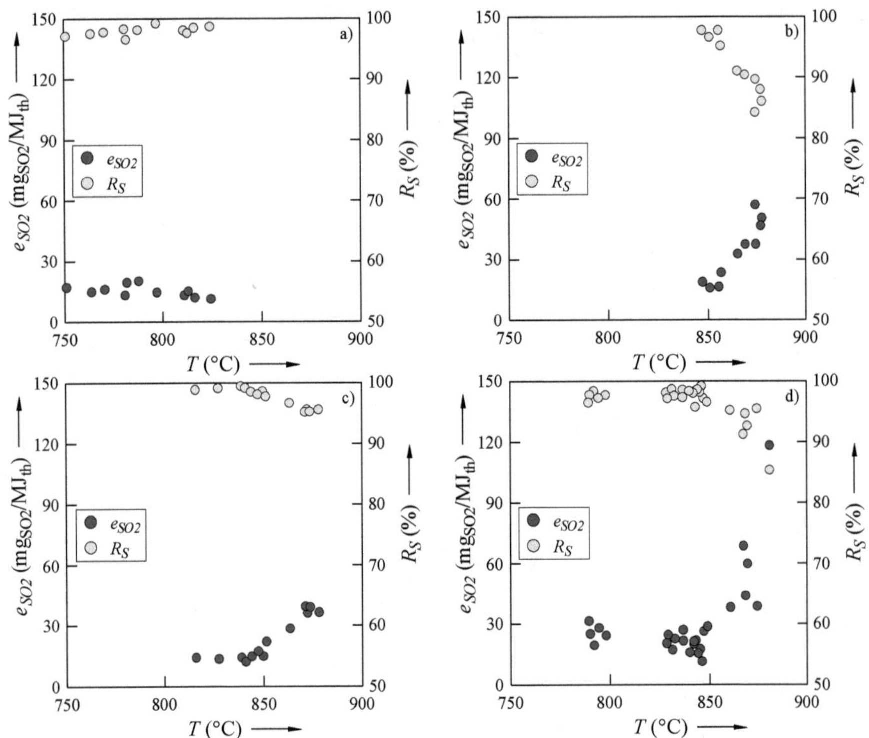

Figure 3-34: Specific SO_2 emission (e_{SO2}) and sulfur retention rate (R_S) as a function of the furnace temperature during TP I (a: SRF I, oxygen-enriched air; b: SRF I, oxyfuel; c: SRF II, oxygen-enriched air; d: SRF II, oxyfuel).

It is known that the ideal temperature for desulfurization during the atmospheric FB combustion of coal using air as oxidation agent is approximately 850 °C [231, 232]. At higher furnace temperatures, the sulfation is limited by the chemical equilibrium. However, this temperature can only be interpreted as a rough estimate as several parameters influence the sulfation of limestone in each particular case. Furthermore, it is questionable whether some minority species of SRF might hinder or positively affect the sulfation pattern, which ultimately influences the ideal temperature for SO_2 absorption. Nevertheless, for both types of SRF, the phenomena related to the formation and/or mitigation of gaseous SO_2 show a similar characteristic. When the furnace temperature ranges below approximately 850 °C, the sulfur retention rate decreases slightly or remains constant along with decreasing temperatures. In that region, the specific SO_2 emissions were below 30 $mg_{SO2}/MJ_{th,LHV}$. Above 850 °C, there is a rapid decrease in the sulfur retention rate, which results in increasing specific SO_2 emissions. For the conditions of Graph a (TP I, SRF I, oxygen-enriched air), this temperature region was not reached in the course of the experimental investigations.

At approximately 875 °C, the lowest sulfur retention rate was nearly 85 % for both types of SRF under oxyfuel conditions. The related specific SO_2 emissions increase to 60 - 65 $mg_{SO2}/MJ_{th,LHV}$. According to the bottom line Figure 3-34 (SRF II), it can be seen, that there is almost no effect of the combustion environment on the sulfur retention rate. For the FB combustion of fossil fuels such as hard coal or lignite, it was found that the oxyfuel combustion environment enhances the sulfation phenomenon [96, 233]. More experimental investigations are required to state whether this might be also the case for the combustion of waste-derived fuels.

In the course of TP II, the volumetric SO_2 was below the detection limit for the majority of test points which results in a sulfur retention rate above 97 %. This seems to be reasonable as to the solid inventory of the furnace exclusively consist of Ca-species and a smaller fraction of fuel ash. Consequently, the molar Ca/S ratio, which was greater than 1,000 in that test period, significantly supports the SO_2 absorption. Due to this reason, the evaluation of SO_2 emissions is limited to results achieved during TP I.

3.10.1.5.4 Hydrogen Chloride

Figure 3-35 shows the specific emissions of HCl (e_{HCl}) and the derived Cl retention rates (R_{Cl}) as a function of the furnace temperature (T) during TP I.

The volumetric concentrations of HCl in the corresponding flue gas stream were in the range of 91 - 131 ppm_v for SRF I and 132 - 210 ppm_v for SRF II in oxygen-enriched air, respectively. Under oxyfuel conditions, significantly higher HCl concentrations were obtained ($y_{HCl,SRF\,I}$: 342 - 496 ppm_v, $y_{HCl,SRF\,II}$: 270 - 667 ppm_v).

While SRF I is combusted in oxygen-enriched air during TP I, the specific emission of HCl remain rather stable between 50 and 75 mg_{HCl}/MJ_{th}. The corresponding chlorine retention rate varies from 74 up to 81 %. Once the combustion environment was switched to oxyfuel atmosphere, the retention rate of chlorine decreases to 18 - 40 %, which causes specific emissions of HCl in the range of 124 - 242 mg_{HCl}/MJ_{th}. Nevertheless, it needs to be noted that both combustion environment tests proceeded in different temperatures ranges, so an additionally influence of the temperature cannot be excluded here. In case of firing SRF II under the aforementioned conditions, the emission formation pattern of HCl show a similar characteristic as for SRF I. Accordingly, the oxyfuel combustion environment is less favorable in terms of Cl retention. Expressed in numbers, this implies a reduction of the average Cl retention from 65.4 to 41.0 %. The specific HCl emissions increase from approximately 167 to 277 mg_{HCl}/MJ_{th}. The negative impact of the oxyfuel combustion environment might be partially explained by the higher volumetric concentration of water vapor, which was found to depress the $CaCl_2$ formation [102].

Even though the chlorine retention rate under oxyfuel conditions is nearly in the same range for both types of SRF, the specific emissions of HCl range up to 350 mg_{HCl}/MJ_{th} while fringing SRF II. This is due to higher specific chlorine content in SRF II in comparison to SRF I ($x_{Cl,spec,SRF\,I}$ = 0.28 g/MJ_{th}, $x_{Cl,spec,SRF\,II}$ = 0.47 g/MJ_{th}). Among the available data points, no distinct influence of the furnace temperature on the HCl emission behavior could be determined.

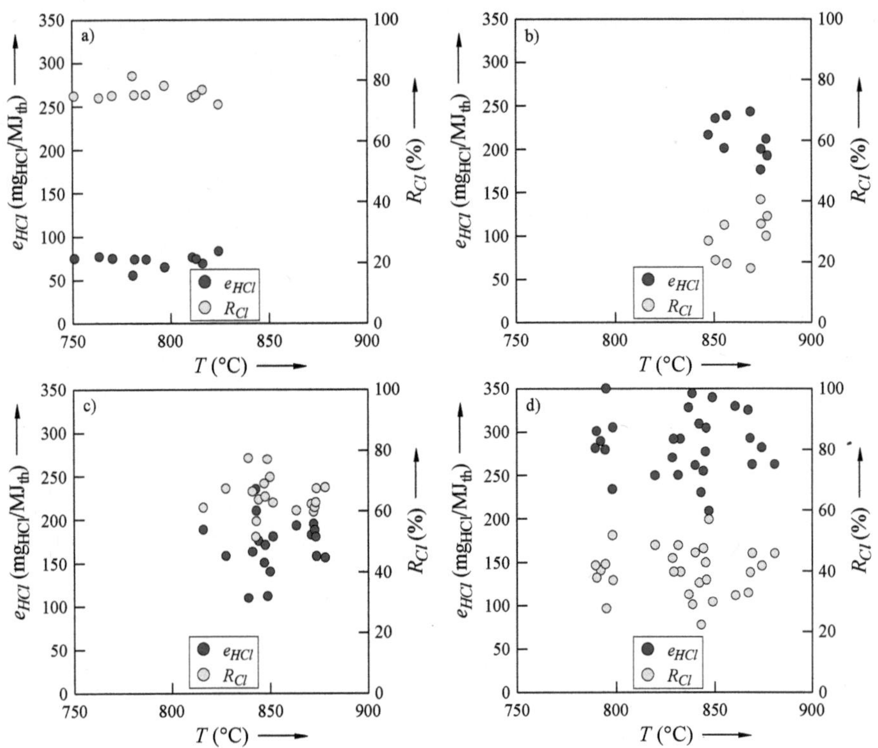

Figure 3-35: Specific HCl emission (e_{HCl}) and chlorine retention rate (R_{Cl}) as a function of the furnace temperature during TP I (a: SRF I, oxygen-enriched air; b: SRF I, oxyfuel; c: SRF II, oxygen-enriched air; d: SRF II, oxyfuel).

Figure 3-36 shows the specific emissions of HCl and the derived Cl retention rates as a function of the riser temperature for the SRF combustion under the boundary conditions of TP II. The volumetric concentrations of HCl in the corresponding flue gas stream were in the range of 4.0 -9.8 ppm$_v$ for SRF I and 50.8 - 72.5 ppm$_v$ for SRF II in oxygen-enriched air, respectively. Under oxyfuel conditions, higher HCl concentrations were obtained ($y_{HCl,SRF\,I}$ = 52.0 - 263 ppm$_v$, $y_{HCl,SRF\,II}$: 41.5 - 194 ppm$_v$).

The chlorine retention rate is significantly enhanced under the combustion conditions in the CaL calciner. This is mainly related to the high fraction of Ca-species in the solid hold up of the calciner, which favors HCl absorption. During oxygen-enriched air combustion of SRF I, the Cl retention rate is close to 100 %, thus the specific emissions at the outlet of the furnace were below 4.1 mg$_{HCl}$/MJ$_{th}$. Even the combustion of SRF II shows a specific emission of HCl below 40 mg$_{HCl}$/MJ$_{th}$ in oxygen-enriched air. Similarly to the results of TP I, the switch to oxyfuel combustion environment causes higher emissions of HCl as consequence of reduced Cl retention. However, the negative effect is less pronounced, as the Cl retention rate is still above 80 % for SRF I and SRF II, respectively.

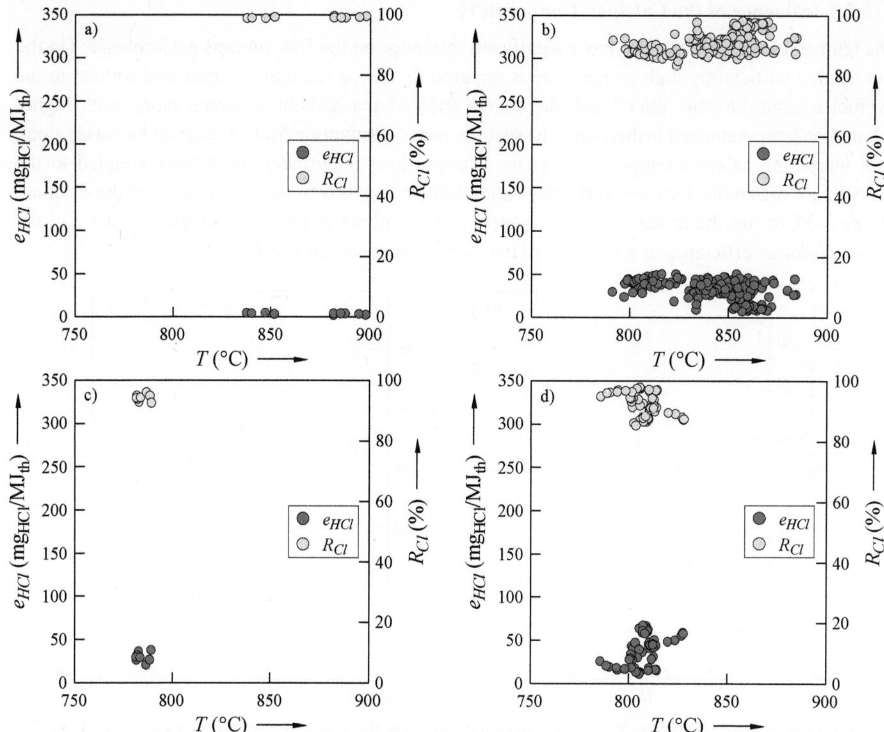

Figure 3-36: Specific HCl emission (e_{HCl}) and chlorine retention rate (R_{Cl}) as a function of the furnace temperature during TP II (a: SRF I, oxygen-enriched air; b: SRF I, oxyfuel; c: SRF II, oxygen-enriched air: d: SRF II, oxyfuel).

3.10.2 Evaluation of Sorbent Regeneration

In the calciner, the partially carbonated solid stream, which is introduced from the carbonator, is regenerated. More precisely, the temperature is raised in order to decompose the contained $CaCO_3$ due to the calcination reaction. The heat required for the sorbent calcination is supplied by the combustion of supplementary fuel. To attain an almost pure CO_2 stream at the outlet of the calciner, the combustion is carried out in an oxyfuel environment, i.e. a mixture of technically pure O_2 and recirculated flue gas is utilized as oxidation agent. Similar to the CO_2 absorption efficiency of the carbonator, one could derive the calciner efficiency, E_{calc}:

$$E_{calc} = 1 - \frac{X_{calc}}{X_{carb}}$$

(3-27)

Here, X_{calc} is the molar fraction of Ca as $CaCO_3$ at the inlet of the carbonator, and X_{carb}, is the molar fraction of Ca as $CaCO_3$ at the outlet of the carbonator. Both numbers are derived after the experiment by solid sample analysis. Similar to the carbonator, several parameters influence the calciner performance in terms of calcination efficiency. In the course of this evaluation, the effect of the average calciner riser temperature, of the calciner CO_2 partial pressure ratio and of the calciner active space time on the sorbent regeneration is investigated.

3.10.2.1 Influence of the Calciner Temperature

The temperature of the calciner has a significant influence on the CaL process performance. On the one hand, a sufficiently high temperature is required to calcine the partly carbonated sorbent in the oxyfuel regime. On the other hand, the deactivation of the sorbent is forced along with higher calcination temperatures. Furthermore, the specific heat consumption for CO_2 capture increases along with increasing calciner temperatures as the sensible heat requirement is directly coupled to the temperature difference between carbonator and calciner and thus to the temperature of the calciner. Figure 3-37 shows the molar carbonate content of the sorbent at the outlet of the calciner and the derived calciner efficiency as a function of the average calciner temperature.

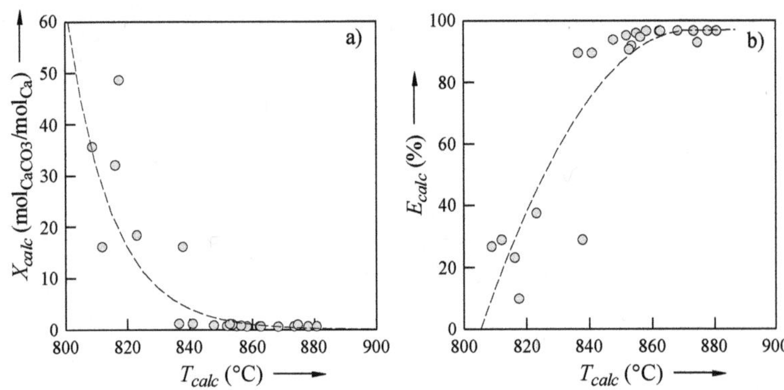

Figure 3-37: Graph a: Molar carbonate content at the outlet of the calciner (X_{calc}), and Graph b: Calciner efficiency (E_{calc}) as a function of the calciner temperature (T_{calc}).

A calciner temperature above 850 °C is sufficient to keep the calcium carbonate content well below 3 mol_{CaCO3}/mol_{Ca}. This results in calciner efficiencies above 90 %. Once the temperature drops below 840 °C, the calciner performances decreases exponentially towards unfavorable values. At an average temperature of 820 °C, the calciner efficiency is below 40 %.

The temperature of the calciner influences the rate of the calcination reaction and the limitation set by the chemical equilibrium of the CaO-$CaCO_3$ system. Therefore, prevailing CO_2 partial pressure conditions needs to be taken into account in the course of the discussion of the calciner performance as well.

3.10.2.2 Calciner CO₂ Partial Pressure Ratio

As noted, the calciner efficiency is mainly dependent on the calciner average temperature and the prevailing CO_2 concentration. These two parameters are directly linked by the chemical equilibrium of the CaO-$CaCO_3$ system. A calciner temperature close to 900 °C is required to ensure a complete sorbent calcination under ideal conditions of an oxy-fired CaL calciner. Ideal conditions in this case are associated with a relatively high CO_2 partial pressure in the flue gas exiting the calciner. At the 1 MW$_{th}$ pilot plant, a dilution of the CO_2 product stream at the outlet of the calciner occurs due to practical reasons. This is associated with the nitrogen flushing of the SRF feeding system, the fluidization of the calciner loop seal and the flushing of the pressure measurement devices. Furthermore, a relatively large oxygen excess was required to ensure the oxidation of the SRF combustible matter at all times (see Chapter 3.10.1). A sufficient sorbent calcination was therefore feasible at calciner temperatures below 900 °C.

In order to take into account the actual CO_2 partial pressure in relation to the calciner temperature, the calciner CO_2 partial pressure ratio, Ω_{calc} is calculated:

$$\Omega_{calc} = \frac{y_{CO2,calc,out}}{y_{CO2,eq}} \tag{3-28}$$

Here, $y_{CO2,calc,out}$ is the CO_2 concentration at the outlet of the calciner, and, $y_{CO2,eq}$ is the CO_2 equilibrium concentration as a function of the calciner temperature [116]. For a sufficient sorbent calcination, the calciner CO_2 partial pressure ratio needs to be < 1, whereas a value > 1 indicates no calcination or even re-carbonation in the calciner. However this evaluation approach is based on the outlet concentration of CO_2. It is worth mentioning that the calciner CO_2 partial pressure ratio is expected to vary over the height of the calciner riser, while taking into account the different stages of SRF combustion and sorbent calcination. Figure 3-38 shows the calciner CO_2 partial pressure ratio as a function of the average calciner temperature for two different ranges of oxygen-to-fuel ratios. Additionally, the critical CO_2 partial pressure threshold of 1 and the minimum calciner temperature for calcination are marked.

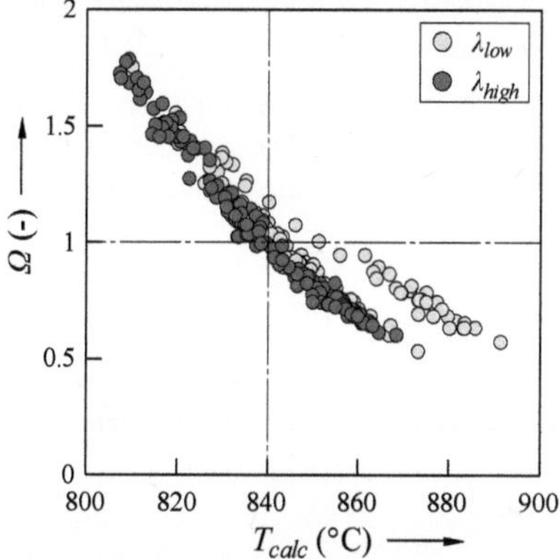

Figure 3-38: Calciner CO_2 partial pressure ratio (Ω_{calc}) as a function of the calciner temperature (T_{calc}) for different oxygen-to-fuel ratios (λ_{low} < 1.3, λ_{high} > 1.3).

Under the experimental conditions, the calciner CO_2 partial pressure ratio varied between 0.53 and 1.75, while the calciner temperatures ranged from 820 to 893 °C. The effect of the CO_2 product stream dilution as a result of the oxygen-to-fuel ratio becomes apparent when comparing the two data sets. In case of a higher dilution (λ_{high}) a lower temperature is sufficient to keep the calciner CO_2 partial pressure ratio in the desired range (< 1). Nevertheless, it is clear, that for full-scale CaL applications, a low oxygen-to-fuel ratio is preferable, as the upgrading requirements in the GPU (i.e. auxiliary power consumption) strongly depend on the oxygen concentration in the CO_2-product stream.

Figure 3-39 shows the molar carbonate content of the sorbent at the outlet of the calciner and the derived calciner efficiency as a function of the calciner CO_2 partial pressure ratio. A calciner CO_2 partial pressure ratio below 1 allows for a calciner efficiency above 90 %. Once the calciner CO_2 partial pressure ratio increases above 1, the calciner efficiency drops rapidly, as the calcination reaction is mainly limited by the chemical equilibrium.

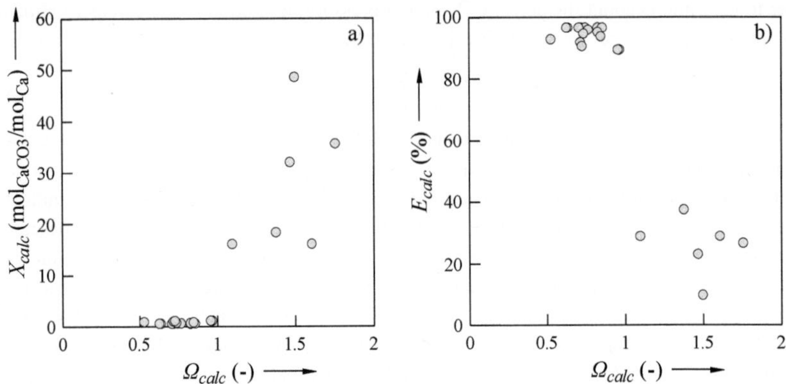

Figure 3-39: Graph a: Molar carbonate content at the outlet of the calciner (X_{calc}) and Graph b: Calciner efficiency (E_{calc}) as a function of calciner CO_2 partial pressure ratio (Ω_{calc}).

3.10.2.3 Calciner Active Space Time

Similarly to the carbonator, the calciner active space time can be applied in the course of calciner performance assessment. In this case, the active fraction refers to the part of carbonated Ca-species within the circulating sorbent stream. The calciner active space time, $\tau_{calc,active}$, is determined as follows:

$$\tau_{calc,active} = \frac{t_{calc}}{X_{carb}} \tag{3-29}$$

Here, t_{calc}, is the mean residence time of particles within the calciner as calculated according to Eq. 3-30 and, X_{carb} is molar carbonated content of the sorbent that enters the calciner.

$$t_{calc} = \frac{n_{Ca,calc}}{F_{Ca}} \tag{3-30}$$

Here, $n_{Ca,calc}$ is the molar inventory of Ca-species in the calciner, and F_{Ca}, is the molar flux of Ca-species transported to the calciner. Figure 3-40 shows the calciner efficiency as a function of the calciner active space for operation points with a calciner CO_2 partial pressure ratio below one.

The calciner active space time ranges from 7.6 to 13.8 min which corresponds to a calciner efficiency between 89.2 and 96.5 %. According to the available data, there is no influence of the calciner active space time on the calcination efficiency. Thus, this metric is mainly dominated by the chemical equilibrium.

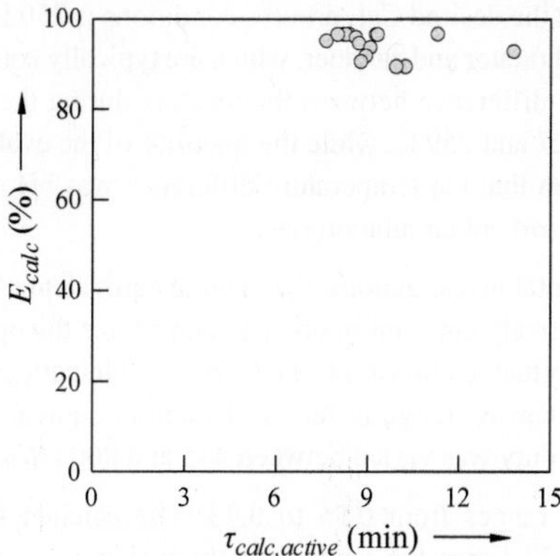

Figure 3-40: Calciner efficiency (E_{calc}) as a function of the calciner active space time ($\tau_{calc,active}$) for data with a calciner CO_2 partial pressure ratio < 1.

3.11 Comprehensive Carbonate Looping Process Performance Assessment

This chapter provides an assessment of the overall CaL process performance. Doing so, the specific requirements of heat, oxygen and limestone make-up for CO_2 capture are derived. These parameters are applicable for the comparison of full-scale and pilot-scale CaL process studies.

3.11.1 Carbonate Looping Process Heat Ratio

The CaL process heat ratio (Eq. 3-11) is governed by process parameters such as the specific sorbent circulation rate and the difference of carbonator and calciner average riser temperature. Additionally, the make-up limestone feeding rate and the conditions of the oxyfuel combustion in the calciner (e.g. oxygen-to-fuel ratio, oxygen inlet concentration) have a minor effect. Figure 3-41 shows the CaL process heat ratio dependent on the difference between calciner and carbonator average riser temperature ($\Delta T_{(calc-carb)}$).

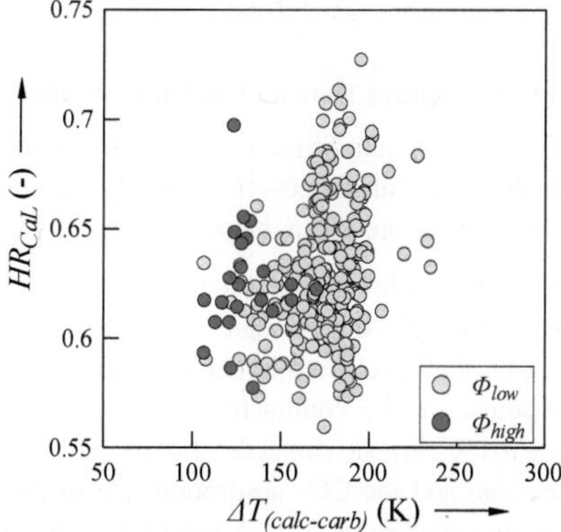

Figure 3-41: CaL process heat ratio (HR_{CaL}) dependent on the temperature difference between calciner and carbonator ($\Delta T_{(calc-carb)}$) for low $\Phi_{low} < 18.04$ and $\Phi_{high} > 18.04$.

The temperature difference at the desired CaL process conditions is 250 K. This is justified due to the operation temperatures of carbonator and calciner, which are typically considered to operate at 650 °C and 900 °C. The temperature difference between the reactors during the course of the experimental investigation was between 107 and 259 K, while the majority of the evaluation points were between 150 and 200 K. It can be seen that the temperature difference was below 175 K for all data points referring to the high specific sorbent circulation rate.

In the course of the experimental investigations, the thermal equivalent of the carbonator (i.e. the flue gas mass flow) was kept relatively constant in order to emphasize the operation of the calciner with particular focus on SRF oxyfuel combustion and sorbent calcination. The carbonator was thus operated within a relatively narrow range in terms of thermal equivalent power (315 - 380 kW_{th}), whereas the calciner thermal duty was varied between 445 and 800 kW_{th}.

The CaL process heat ratio ranges from 0.55 to 0.73. The calciner thus needs to be fueled by approximately 1.22 up to 2.33 times the equivalent thermal duty of the carbonator for flue gas decarbonization associated with the conditions of the 1 MW_{th} CaL pilot plant. This number is significantly higher in contrast to the aforementioned full-scale CaL process integration studies, wherein the CaL process heat ratio ranges from 0.38 to 0.57 (see Table 2-5). This large distinction is mainly due to the semi-industrial characteristics of the 1 MW_{th} CaL pilot plant. More precisely, specific heat losses from the solid looping cycle (i.e. CFB reactors, coupling components) are relatively large in comparison to large-scale applications because of the high specific surface area of the reactor system. Furthermore, the CaL process heat ratio is worse, as the 1 MW_{th} calciner is not specifically designed for the type of fuel that is burnt. Therefore, the oxygen-to-fuel-ratio was relatively high, which further raised the sensible heat demand of the calciner (see Chapter 3.10.1).

A lower CaL process heat ratio is associated with a lower temperature difference for both the high and low specific circulation rate data sets. This is justified by the fact that less sorbent needs to be heated in the calciner per carbonation-calcination reaction cycle. Nevertheless, this evaluation parameter does not consider the effects of the reactor temperature on the CO_2 capture performance. For example, if the calciner temperature is relatively low, the CaL process heat ratio may be advantageous, but in parallel the total CO_2 capture rate might be limited due to insufficient sorbent calcination.

3.11.2 Dependency of Total CO_2 Capture Rate on Carbonator Absorption Efficiency

Figure 3-42 shows the total CO_2 capture rate of the CaL process dependent on the carbonator CO_2 absorption efficiency for three different CaL process heat ratios. Additionally, the trend lines for the low and the high CaL process heat ratios are given, respectively.

For all CaL process heat ratios, there is a direct linkage between the CO_2 absorption rate and the total CO_2 capture rate. The total amount of CO_2 that is captured by the CaL process consists of CO_2 that is absorbed in the carbonator, CO_2 that is released by the first calcination of the make-up limestone and CO_2 that is formed in the course of SRF combustion (see Eq. 3,-6). The latter two fractions are captured nearly completely due to the intrinsic characteristics of the oxyfuel calciner. The correlation between the total CO_2 capture rate and the CO_2 absorption rate of the carbonator shows a lower gradient in case of a higher CaL process heat ratio, as more CO_2 is additionally released in the calciner per CO_2 that is absorbed in the carbonator. In case of a carbonator CO_2 absorption rate of 75 % the total CO_2 capture rate yields to 90.6, or 89.1 % in case of the high or low CaL process heat ratio, respectively. This relation could also be expressed by the specific heat demand for CO_2 capture, as described in the following section.

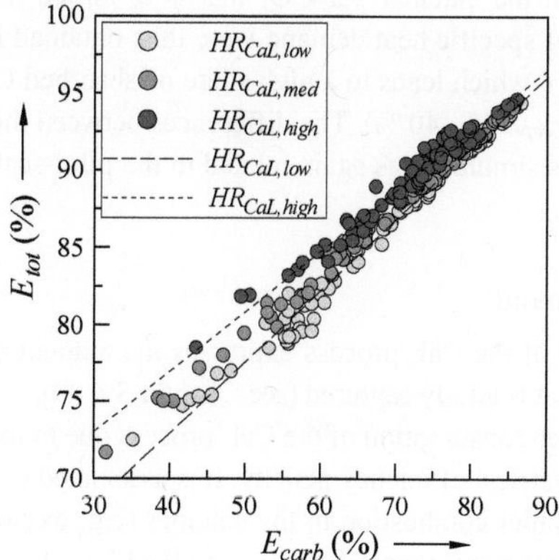

Figure 3-42: Total CO_2 capture rate (E_{tot}) dependent on the CO_2 absorption efficiency in the carbonator (E_{carb}) for different CaL process heat ratios ($HR_{CaL,low} < 0.62$, $HR_{CaL,med} = 0.62 - 0.64$, $HR_{CaL,high} > 0.64$).

3.11.3 Specific Heat Demand

The specific heat demand of the CaL process describes the heat released by the oxyfuel combustion of supplementary fuel in the calciner per amount of CO_2 that is totally captured totally (see Chapter 3.4, Eq. 3-13). This parameter directly affects the oxygen demand and therefore the auxiliary power of the ASU in the directly heated CaL process layout. Low numbers are favorable here. According to the results of process simulations for large-scale CaL applications, this parameter ranges from 3.65 to 5.10 MJ$_{th}$/kg$_{CO2}$ [202, 203].

Figure 3-43 shows the specific heat demand of the CaL process dependent on the CaL process heat ratio (a) and on the share of CO_2 absorbed in the carbonator (b) based on the experimental data.

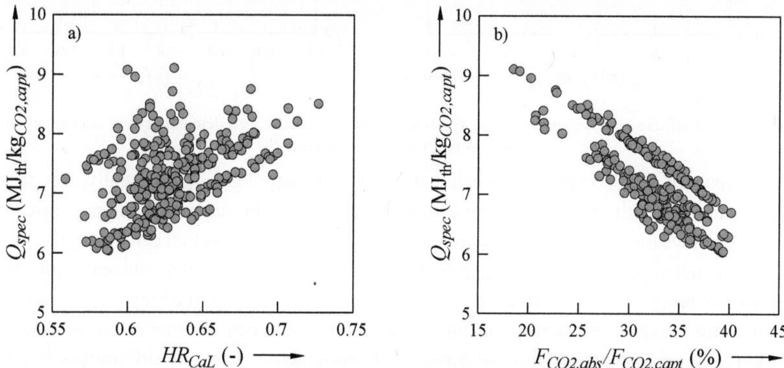

Figure 3-43: Graph a: Specific heat demand (Q_{spec}) as a function of the CaL process heat ratio (HR_{CaL}), Graph b: Specific heat demand as a function of the share of CO_2 being absorbed in the carbonator ($F_{CO2,abs}/F_{CO2,capt}$).

The specific heat demand varied between 6.04 and 9.09 MJ$_{th}$/kg$_{CO2,capt}$. The influences of the CaL process heat ratio and the share of CO_2, which is being absorbed in the carbonator among the total

CO_2 that is captured on the specific heat demand for CO_2 capture are inversely pronounced. This is due to the fact that a high CaL process heat ratio is inherently associated with a high amount of CO_2 being additionally released in the calciner per CO_2 that is absorbed in the carbonator. The most favorable numbers in terms of specific heat demand were thus obtained in case of low CaL process heat ratio (HR_{CaL}: 0.55 - 0.60), which leads to a high share of absorbed CO_2 among to the total CO_2 that is captured ($F_{CO2,abs}/F_{CO2,capt}$: 35 - 40 %). The differences between the experimental data and the results from full-scale process simulation is again related to the pilot-scale characteristics of the test facility.

3.11.4 Specific Oxygen Demand

The specific oxygen demand of the CaL process expresses the amount of oxygen that is fed to the calciner per amount of CO_2 that is totally captured (see Chapter 3.4, Eq. 3-14). This number is directly assigned to the auxiliary power consumption of the CaL process due to the operation of the ASU and therefore related to the net electrical efficiency penalty. It is associated with the specific heat demand and the conditions of the oxyfuel combustion in the calciner (e.g. oxygen-to-fuel ratio). For large-scale CaL applications, this parameter ranges from 0.34 to 0.44 $kg_{O2}/kg_{CO2,capt}$ [192, 202].

Figure 3-44 shows the specific oxygen demand dependent on the specific heat demand (Graph a) and on the oxygen-to-fuel ratio of the calciner (Graph b).

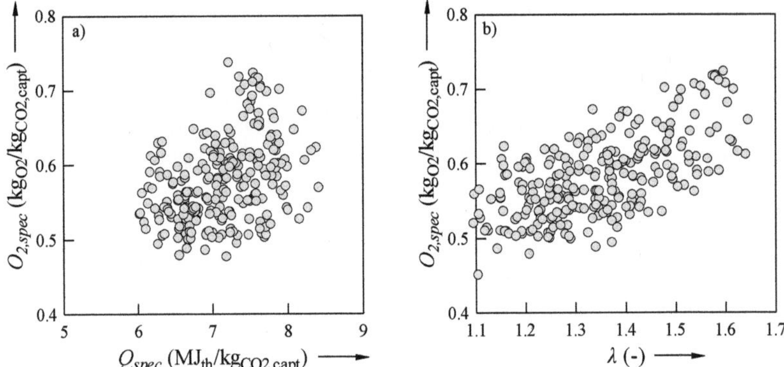

Figure 3-44: Specific oxygen demand ($O_{2,spec}$) as a function of the specific heat demand (a) and dependent on the oxygen-to-fuel ratio in the calciner (b).

The specific oxygen demand varied between 0.465 and 0.733 $kg_{O2}/kg_{CO2,capt}$. Trivially, this parameter is directly associated with the specific heat demand so that a low thermal requirement per quantity of captured CO_2 allows for a low specific oxygen demand. In case of the oxyfuel calciner, both quantities are physically decoupled by the oxygen-to-fuel ratio, the oxygen inlet concentration and the combustion efficiency. Figure 3-44b clearly indicates the positive effect of a low oxygen-to-fuel-ratio on the specific oxygen demand. An unfavorable high oxygen consumption of more than 0.7 $kg_{O2}/kg_{CO2,capt}$ is associated with oxygen-to-fuel ratios above 1.5. The differences in terms of specific oxygen consumption between large-scale CaL process studies and the data obtained from 1 MW_{th} pilot testing are mainly related to the fact that the test facility is not specifically optimized for the combustion of SRF under CaL process conditions.

3.11.5 Specific Make-up Requirement

As stated in Chapter 3.9.2, a minimum make-up feeding rate is required to achieve the desired CO_2 absorption rate in the carbonator. Dependent on the boundary conditions of the CaL process (e.g. type of supplementary fuel, composition of the flue gas) the make-up is required to maintain a reasonable CO_2 carrying capacity of the sorbent or to allow for a higher purge rate, as it is required in case of firing SRF. Nevertheless, the make-up limestone feed needs to be as low as possible to reduce the material efforts and costs related to the operation of the CO_2 capture process.

Figure 3-45 shows the specific make-up requirement dependent on the nominalized carbonator CO_2 absorption efficiency for three different specific make-up rates.

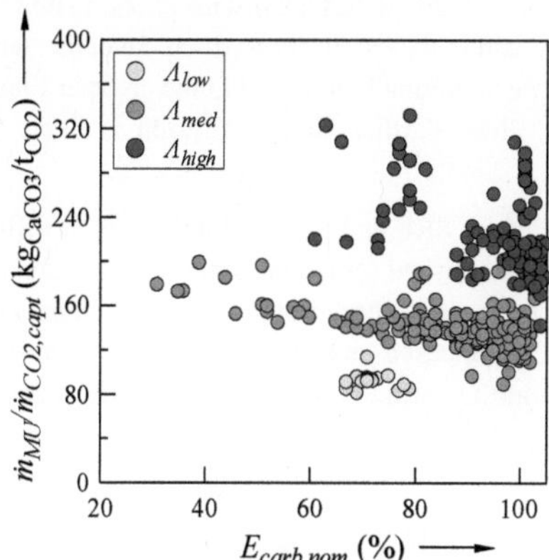

Figure 3-45: Specific make-up requirement ($\dot{m}_{MU}/\dot{m}_{CO2,capt}$) dependent on the nominalized carbonator CO_2 absorption efficiency ($E_{carb,nom}$) for three different specific make-up rates ($\Lambda_{low} < 0.100$; $\Lambda_{med} = 0.100 - 0.175$; $\Lambda_{high} > 0.175$).

The specific make-up requirement ranges from 81 kg$_{CaCO3}$/t$_{CO2,capt}$ (Λ_{low}) up to 330 kg$_{CaCO3}$/t$_{CO2,capt}$ (Λ_{high}). For all specific make-up feeding rate data sets, the specific make-up requirement decreases once more CO_2 is absorbed in the carbonator, which is associated with higher nominalized carbonator CO_2 absorption efficiencies. Furthermore, it comes apparent that lower quantities of fresh make-up limestone are associated with a low specific make-up requirement and only moderate carbonator CO_2 absorption rates as in the case of data points associated with Λ_{low}.

3.11.6 The Fate of Chlorine

Understanding the fate of chlorine within the CaL system is crucial for an appropriate design of subsequent heat recovery systems and gas cleaning units. Any form of chlorine-induced corrosion of boiler materials is of major concern here [29, 234, 235]. Chlorine might be introduced to the CaL process via the flue gas to be decarbonized in the carbonator in the form of HCl and via the various waste fractions contained in SRF. During the experimental investigations, the chlorine feed was limited to the latter quantity.

During the oxyfuel combustion in the calciner, chlorine is released to the gas phase via HCl and partially absorbed by the solid phase. Part of the solid phase bound chlorine is transported to the carbonator by the circulating sorbent. Chlorine is thus released by the CaL process via gaseous and solid streams from the carbonator and from the calciner respectively. Table 3-14 summarizes the range of the total mass flow of Cl-containing solid material streams in the CaL process and the range of the corresponding Cl mass fractions in the associated solid samples.

Table 3-14: Chlorine-containing solid material streams in the CaL process (BA: Bottom ash, FA: Fly ash).

Parameter	Unit	BA_{Carb}	BA_{Calc}	[12] FA_{Carb}	FA_{Calc}	$Carb_{in}$
Total mass flow, $\dot{m}$	kg/h	2.5 - 38	5.0 - 58	1.0 - 21	5.8 - 44	1,800 - 3,700
Cl mass fraction, x_{Cl}	wt.%	0.20 - 0.32	0.21 - 0.36	1.6	2.2 - 3.3	0.16 - 0.75

The chemical solid analysis shows that the chlorine concentration is highest in the calciner fly ash (x_{Cl}: 2.2 - 3.3 wt.%), while carbonator fly ash shows a lower value in the same order of magnitued. In contrast to that, the chlorine mass fraction in the bottom ashes discharged from carbonator and calciner are significantly lower (x_{Cl}: 0.20 - 0.36 wt.%). This phenomenon is mainly caused by the fine Ca-particles present in the fly ashes ($d_{p,50} < 50$ µm) which favor the absorption of gaseous HCl. Furthermore, fly ash particles are additionally exposed to the calciner flue gases during the course of heat removal and in the bag filter. Further HCl absorption is likely to occur in these process sections [236].

Based on the chlorine introduced by SRF and the solid and gaseous effluent streams from the CaL process, a chlorine balance was established for a representative operation point (see Figure 3-46). For this balance, it is assumed that the solid phase bound chlorine remains constant in the carbonator. This implies that neither Cl is released nor additional HCl is absorbed by the solid phase. Additionally, the HCl absorption that is likely to occur at the filter cake of the calciner bag filter is neglected.

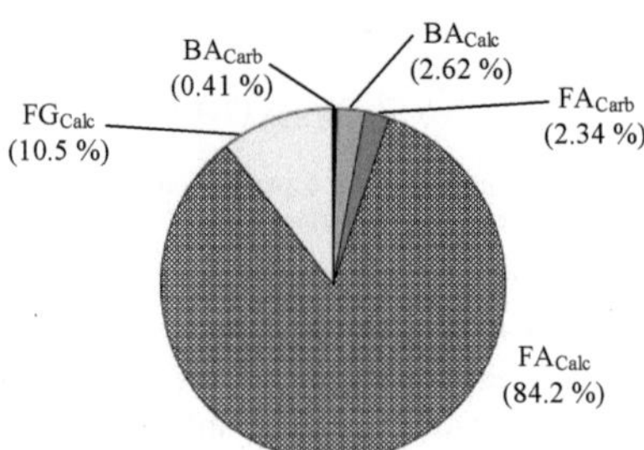

Figure 3-46: The fate of chlorine for a representative operation point.

According to the chlorine balance, 15.4 mol/h are introduced to the calciner via SRF, while 19.1 mol/h are released according to the considered effluent streams. There is a deviation of approximately 23 % as more chlorine is released. This might be due to several reasons, such as the fluctuation of the actual chlorine mass fraction of the SRF or the fact that the extracted solid samples represent only a minor share compared to the corresponding total mass flows in the CaL system. Furthermore, it is likely that HCl absorption from the calciner flue gas continues in the course of the flue gas path downstream the calciner cyclone. More precisely, a significant share of HCl is absorbed by the filter cake which built up at the bag filter of the calciner. Here, the HCl containing flue gas passes through a layer of highly active Ca-species in a temperature range of 175 - 225 °C. These

---

[12] Only one representative sample was available for the carbonator fly ash.

conditions are known to allow for HCl absorption [237]. Due to the fact that the HCl concentration in the flue gas was measured upstream of the calciner bag filter, the HCl effluent in the calciner flue gas tends to be overestimated.

The fly ash of the calciner represents the major chlorine effluent from the CaL process with a fraction of 84.2 %. The calciner flue gas represents another relevant effluent (10.5 %). The chlorine quantities contained in the fly ash from carbonator and in the bottom ashes from both reactors (< 3 %) are of lesser importance, respectively.

3.12 Experimental Summary

In the experimental section, the utilization of SRF in the CaL process is evaluated thoroughly. The experimental investigations were carried out at the 1 MW_{th} CaL pilot plant at Technische Universität Darmstadt in the course of two consecutive test campaigns. In addition to the novel type of supplementary fuel, the CO_2 concentration of the flue gas to be decarbonized in the carbonator was adapted according to values of a typical WtE plant flue gas fueled by MSW. Both experimental boundary conditions aim at the application of the CaL process in the framework of WtE plants.

By more than 230 hours of representative CaL operation conditions at the 1 MW_{th} CaL pilot plant, the continuous CO_2 capture was successfully demonstrated. Carbonator CO_2 absorption rates close to the chemical equilibrium and total CO_2 capture rates over 90 % were experimentally proven. Furthermore, it was shown that a stable pilot plant operation is feasible despite the highly heterogeneous nature of SRF. Attention needs to be paid to distribution and accumulation of coarse inert ash fraction in the solid looping system. Throughout all test points, the carbonator was operated without active reactor cooling (i.e. the available cooling tubes were not inserted into the reactor). This is feasible as less CO_2 is present in the flue gas to be decarbonized. The evaluation of CaL process performance is based on the closure of carbon balances for the carbonator and for the total CaL system including carbonator and calciner. It was found that a carbonator active space time of 60 s is required to achieve a nominalized CO_2 absorption efficiency above 80 %. The analysis of the calciner reveals that a calcination temperature of 840 °C is sufficient for a calcination efficiency above 90 %, as the CO_2 partial pressure in the riser is reduced by the presence of water vapor, nitrogen and oxygen. On the basis of the numerous solid samples, the abrasion coefficient of the circulating solid was determined. It was proven that the sorbent particles with a carbonate product layer show a higher resistance to mechanical attrition forces in comparison to the fully-calcined particles. The activity of selected samples was determined by means of TGA analysis. The activity ranges from 0.041 to 0.102 mol_{CaCO3}/mol_{Ca}, while a higher make-up feed and a lower calcination temperature enhance sorbent activity.

In addition to the evaluation of sorbent regeneration in the calciner, the FB combustion of SRF was assessed in terms of major gaseous emission pollutants, namely CO, NO, SO_2 and HCl. Here, the combustion in air, oxygen-enriched air and in a real oxyfuel environment is evaluated for the CaL calciner and for the combustion in a stand-alone CFB with a silicate sand solid inventory. For both types of SRF, the combustion in four distinguished regimes is thus assessed. The overall level of specific CO emissions was relatively high throughout all experimental investigations, which indicates a poor fuel burnout. This might be associated with a fluctuating fuel mass flow and a poor mixing of volatile species that move upwards in the wall region of riser. Both phenomena can be significantly improved by an appropriate design at a large-scale unit. The formation tendency of NO is in accordance with the fundamentals related to the combustion of conventional solid fuels. Accordingly, a higher oxygen-to-fuel ratio and a higher nitrogen fraction in the fuel support the formation of NO. Contrary to that, an oxyfuel environment lowers the specific emission of NO. The specific emissions of SO_2 vary between 20 and 70 mg_{SO2}/MJ_{th}, while the ideal temperature for sulfation of Ca-species was found to be approximately 850 °C. However, the specific emissions of SO_2 during the SRF combustion in the CaL calciner were very low and close to the detection limit of the gas analyzer throughout the experimental investigations. The emission characteristics of HCl are also greatly influenced by the properties of the solid phase. Accordingly, Cl retention rates above 80 % were achieved during the course of oxyfuel combustion in the CaL calciner. However, the Cl retention decreases to 18 - 58 % during oxyfuel combustion in a silicate sand CFB.

4 Carbonate Looping Process Modelling

The CaL technology has proven its feasibility in semi-industrial scale by pilot testing in various test rigs all over the world. For the further development, comprehensive models are required for the prediction of process characteristics in terms of heat and mass flows. The appropriate modelling of the CO_2 absorption in the carbonator and the calculation of sorbent activity are most crucial to the determination of overall CaL process performance. This section describes and validates a detailed process model for the CaL process based on the experimental data of 1 MW_{th} pilot testing.

4.1 Model Description

The process model is based on the ASPEN PLUS™ flowsheet simulation environment. A detailed calculation sequence that emphasizes the CO_2 absorption in the carbonator is implemented in the flowsheet simulation via a FORTRAN interface. The model was initially proposed by Romano [180]. Accordingly, the main parts of the calculation sequence are as follows:

- CFB carbonator hydrodynamics

- CO_2 absorption efficiency based on the molar conversion

- CO_2 absorption efficiency based on the material balance

The calculation sequence is graphically summarized in Figure 4-1. As first part of the calculation procedure, the hydrodynamics of the CFB carbonator are determined. This particularly implies the height of lean and dense region as gas-solid mass transfer is different in both regions. Thereafter, the CO_2 absorption efficiency is determined by means of two different approaches. The first approach takes into account the solid phase related properties (e.g. reaction age distribution of the sorbent). The second approach is related to the material balance of the carbonator and takes into account the overall reaction rate of the carbonation reaction. The main idea of the model is the iterative comparison of the results from both approaches. Based on a hypothetical initial start value of the CO_2 concentration, the calculation sequence is iteratively repeated until a sufficient agreement between both approaches is reached.

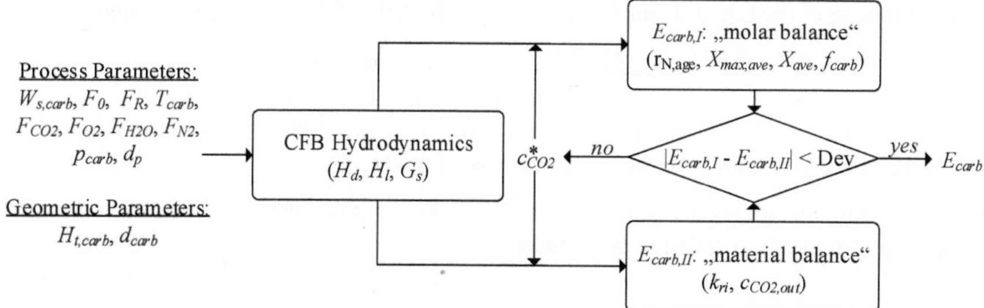

Figure 4-1: Sequence for the calculation of CO_2 absorption in the carbonator.

The process model is based on the following assumptions:

- Even temperature and superficial gas velocity distribution in the carbonator riser.

- No mass transfer limitation by the gas phase, perfect particle mixing.

- Uniform particle size of the particles. No particle size distribution is considered.

4.1.1   CFB Carbonator Hydrodynamics

The solid distribution within the carbonator riser is calculated with an empirical, one-dimensional model for fast fluidization from Kunii and Levenspiel [189, 238]. Here, the carbonator riser is divided into two hydrodynamic regions. The dense region in the lower part of the reactor and the upper, lean region. In the dense region, a constant particle volume concentration, ε_{sd}, is assumed, whereas the solid population in the lean region decreases exponentially according to the decay constant, a. The decay constant is dependent on the superficial gas velocity, u_o, such that $au_o = 3$. The volume fraction of solids at the riser outlet, ε_{se}, is calculated according to Eq. 4-1.

$$\varepsilon_{se} = \varepsilon_s^* + (\varepsilon_{sd} - \varepsilon_s^*)\varepsilon^{-aH_l} \tag{4-1}$$

Here, H_l is the height of the lean region. The saturation solid carrying capacity of the gas, ε_s^*, is calculated according to Eq. 4-2, where G_s^* is the saturated mass flow of solids, u_t is the terminal velocity of the particles, and ρ_s is the mass density of the solid.

$$\varepsilon_s^* = \frac{G_s^*}{(u_0 - u_t)\rho_s} \tag{4-2}$$

The saturated mass flux of solids, G_s^*, is calculated with an empirical approach of Geldart et al. [238]:

$$G_s^* = 23.7\rho_g u_0 e^{-\frac{5.4u_t}{u_0}} \tag{4-3}$$

The terminal velocity of a solid particle is derived from fluid mechanics (see Chapter 2.3).

The total inventory of the carbonator, W_t, consists of the sum of the solid mass in the dense, W_d, and in the lean region, W_l, respectively.

$$W_t = W_d + W_l = A_t\rho_s H_d \varepsilon_{sd} + A_t\rho_s H_l f_l \tag{4-4}$$

Here, f_l states the average volume concentration of solids in the lean region:

$$f_l = \varepsilon_s^* + \frac{\varepsilon_{sd} - \varepsilon_{se}}{H_l} \tag{4-5}$$

By the combining of the Eqs. 4-4 and 4-5, it is possible to derive the following expression:

$$\frac{W_t}{A_t\rho_s} = \frac{\varepsilon_{sd} - \varepsilon_{se}}{a} + H_t\varepsilon_{sd} - H_l(\varepsilon_{sd} - \varepsilon_s^*) \tag{4-6}$$

The height of the lean region and the volume fraction of solids at the riser outlet is subsequently calculated by the application of the Newton-Raphson method [240].

Table 4-1 summarizes the model parameters for the hydrodynamic calculation, as well as the range of values that is typically considered in literature.

Table 4-1: Model parameter for the hydrodynamic calculation [83, 189].

Symbol	Unit	Range	Value
c_{sd}	-	0.06-0.22	0.16
ε_{sw}	-	0.4-0.6	0.5
ε_{sc}	-	~0.01	0.01
ε^{*}	-	~0.01	0.01
au_0	-	2-7	3.0

4.1.2 CO₂ Absorption Efficiency Based on the Molar Conversion

The CO_2 carrying capacity of the sorbent is determined depending on the deactivation constant, k_0, and the residual sorbent conversion capacity $X_{r,0}$, as proposed by Grasa and Abanades [123].

$$X_{max,N} = \frac{1}{\frac{1}{1-X_{r,0}}+k_0 N} + X_{r,0} \tag{4-7}$$

In this study, the initial values chosen for the deactivation constant and the residual conversion capacity without sulfation are 0.52 and 0.075 respectively. N represents the number of complete calcination-carbonation cycles.

The degree of sulfation of the sorbent has a significant influence on the evolution of the CO_2 carrying capacity. The related mechanisms of the sorbent deactivation are introduced in Chapter 2.5.4. In order to account for the sulfur induced sorbent deactivation, the deactivation constant and residual conversion capacity are calculated dependent on the molar sulfation level of the sorbent. For this purpose, experimental data from Grasa et al. [181] are applied to Eqs. 4-8 to 4-10:

$$k = k_{r,0} * (1 + 0.2962 * x_{CaSO4}) \tag{4-8}$$

$$X = X_{r,0} * (1 - 1.1536 * x_{CaSO4}) \quad for \; x_{CaSO4} \leq 0.5 \tag{4-9}$$

$$X = X_{r,0} * (0.4230 - 3.076 * x_{CaSO4}) \quad for \; x_{CaSO4} > 0.5 \tag{4-10}$$

The calculation of the CO_2 absorption efficiency is conducted in terms of the sorbent molar conversion. This approach takes into account the average maximum sorbent conversion, $X_{max,ave}$, of the sorbent particles:

$$X_{max,ave} = \sum_{N_{age}=1}^{+\infty} r_{N,age} X_{max,N} \tag{4-11}$$

Here, $r_{N,age}$, expresses the fraction of particles that have experienced N_{age} full carbonation-calcination cycles. It is dependent on the real level of carbonation, f_{carb}, and calcination, f_{calc} [180]:

$$r_{N,age} = \frac{\left(\frac{F_0(1-f_{calc})}{F_0+F_R f_{calc}} + \frac{F_0}{F_R}\right) f_{carb}^{N_{age}-1} f_{calc}^{N_{age}}}{(\frac{F_0}{F_R} + f_{carb} f_{calc})^{N_{age}}} \tag{4-12}$$

The average conversion degree of the particles, X_{ave}, is determined according to Eq 4-13. The average conversion of particles with a residence time larger than t_{lim} is equal to the maximum conversion at the end of the fast carbonation reaction stage, $X_{max,N}$. For particles having a residence time smaller than t_{lim}, the average conversion is a function of a given CO_2 concentration, c_{CO2}^*, and the number of complete calcination-carbonation cycles.

$$X_{ave} = \sum_{N_{age}=1}^{+\infty} r_{N,age} \left(\int_0^{t_{lim}} f(t) \cdot X(t, N_{age}, c_{CO2}^*) dt + \int_{t_{lim}}^{\infty} f(t) \cdot X_{max,N} dt \right) \tag{4-13}$$

The CO_2 absorption efficiency according to the sorbent molar conversion, $E_{carb,I}$ is then obtained by Eq. 4-14 and the given molar flows of CO_2 (F_{CO2}) and CaO (F_R) entering the carbonator.

$$E_{carb,I} = \frac{F_R X_{ave}}{F_{CO2}} \tag{4-14}$$

4.1.3 CO_2 Absorption Efficiency Based on the Material Balance

The calculation sequence for the CO_2 absorption efficiency dependent on the material balance, $E_{carb,II}$ is based on the overall carbonation reaction rate constant, k_{ri}, according to Eq. 4-15 [185].

$$k_{ri} = \frac{\rho_s}{M_s} \sum_{N_{age}=1}^{+\infty} r_{N,age} \left(\int_0^{t_{lim}} f(t) k_s S_N [1 - X(t, N_{age}, c_{CO2}^*)]^{2/3} \right) dt \tag{4-15}$$

Here, k_s, is the intrinsic kinetic constant of the carbonation reaction, set as $6.05 * 10^{-10}$ $m^4/mols$, S_N is the specific surface area of CaO particles having N_{age} complete calcination-carbonation cycles, and c_{CO2}^*, states the given CO_2 concentration as fitting parameter. The amount of CO_2 being absorbed in the carbonator is then calculated according to the Kunii and Levenspiel model for fast fluidized bed reactors [238].

The overall carbonator CO_2 absorption efficiency according to the material balance is finally derived from Eq. 4-16.

$$E_{carb,II} = \frac{F_{CO2} - \dot{V}_{g,out} * c_{CO2,out}}{F_{CO2}} \tag{4-16}$$

Here, $\dot{V}_{g,out}$, is the volumetric flow rate of flue gas at the outlet of the carbonator, and $c_{CO2,out}$ is the CO_2 concentration at the carbonator outlet. If the difference between $E_{carb,I}$ and $E_{carb,II}$ is reasonably small (< 0.005), the average value of both numbers is applied. Otherwise, the initial start value of c_{CO2}^* is adjusted iteratively.

4.2   Model Validation

The CaL process model is validated by the experimental data from the 1 MW_{th} pilot plant (see Chapter 3). The experimental input parameters cover the properties of the gaseous and solid material flows that are introduced into the carbonator. Furthermore, the carbonator average temperature and the carbonator solid inventory are relevant for process modelling.

Table 4-2 summarizes the experimental input parameters which are applied in the course of model validation. The corresponding values for 20 validation points are given in Table A-3 (Appendix).

Table 4-2: Experimental input parameter for process model validation.

Parameter	Symbol	Unit
Molar make-up feed	F_0	mol/s
Circulating sorbent	$\dot{m}_{sorbent}$	kg/h
$CaSO_4$ mass fraction	x_{CaSO4}	wt.%
$CaCO_3$ mass fraction	x_{CaCO3}	wt.%
CaO mass fraction	x_{CaO}	wt.%
Ash mass fraction	x_{Ash}	wt.%
Carbonator solid inventory	W_{carb}	kg
Carbonator temperature	T_{carb}	°C
Molar H_2O flow to the carbonator	F_{H2O}	mol/s
Molar O_2 flow to the carbonator	F_{O2}	mol/s
Molar N_2 flow to the carbonator	F_{N2}	mol/s
Molar CO_2 flow to the carbonator	F_{CO2}	mol/s

Figure 4-2 compares the modelled CO_2 absorption efficiencies ($E_{carb,sim}$) of the carbonator with those obtained from the experiment ($E_{carb,exp}$). Additionally, the relative deviation between experimental data and the process model results is given (ΔE_{carb}).

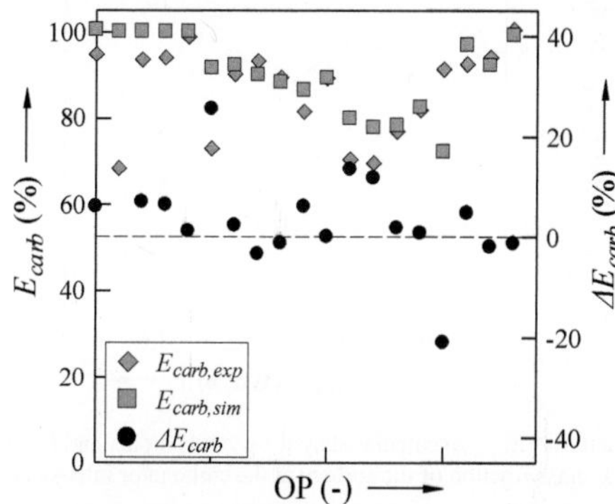

Figure 4-2: Comparison of the CO_2 absorption efficiency derived from the experiment ($E_{carb,exp}$) and from the process model ($E_{carb,sim}$) and relative deviation (ΔE_{carb}).

Overall, the results are in reasonable accordance with the experimental data. The CO_2 absorption efficiency for 75 % of the validation points is reproduced by +/- 10 % related to the experimental data. Generally, the process model tends towards an overestimation of the carbonator CO_2 absorption performance, which might be caused by several reasons. One explanation is related to the deactivation mechanism of the sorbent, which strongly depends on the operation conditions during calcination (e.g. temperature, reaction time). The effects of different calcination conditions on the sorbent characteristics were not taken into account by the process model. It was already proven that the calciner temperature directly influences the sorbent activity (see Chapter 3.7.4). Moreover, the influence of chlorine species and water vapor on the pore structure of the solid phase are also not covered by the process model.

It is important to note that most of the validation points were close the chemical equilibrium conditions. Thus, even if the process model underestimates the deactivation mechanisms, the CO_2 absorption will be in reasonable accordance, as the limiting effect of the chemical equilibrium is well implemented by the equation of Gracia Labiano et al. [116]. Furthermore, the assumption of isothermal conditions in the carbonator might distort the modelling results. Even though the carbonator temperature profile is relatively homogenous (see Chapter 3.6.1), the effect of an additional CO_2 absorption in the colder top region of the reactor is not covered by the process model. Doing so, the CO_2 absorption efficiency would increase even further. Nevertheless, based on the reasonable agreement of Figure 4-2 the process model is suitable to be applied in the further course of process modelling.

In order to further assess the results of the process model by experimental data, a comparison of the maximum average CO_2 carrying capacity of the sorbent is carried out. The experimental data is derived from TGA analysis (see Chapter 3.7.4), the modeling data is derived from Eq. 4-11.

Figure 4-3 shows both data sets as a function of the sulfur mass fraction of the sorbent at the inlet to the carbonator. The molar sulfation level is known to be a crucial parameter in the process of sorbent deactivation.

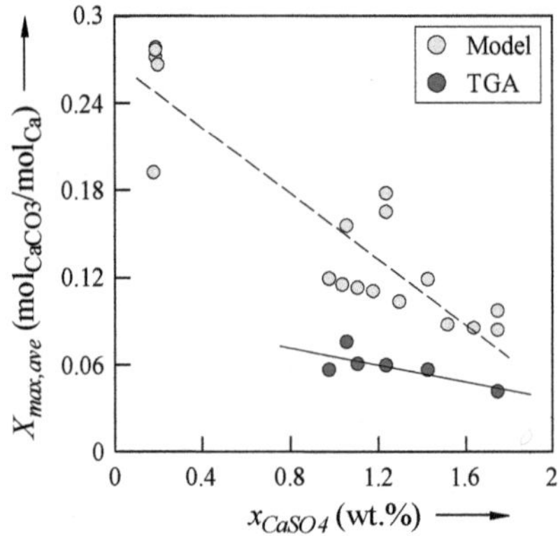

Figure 4-3: Maximum sorbent activity ($X_{max,ave}$) calculated by the process model and by TGA analysis dependent on the $CaSO_4$ mass fraction of the sorbent at the carbonator inlet (x_{CaSO4}).

The sorbent activity according to the process model is qualitatively in reasonable agreement with the experimental data. Accordingly, the calculated maximum average CO_2 carrying capacity decreases by approximately 70 % between the low and high sulfated sorbent ($x_{CaSO4,low}$ = 0.19 wt.%, $x_{CaSO4,high}$ = 1.75 wt.%), respectively. In comparison, the maximum average CO_2 carrying capacity derived by TGA analysis decreases by 45 % between the low and the high sulfated sorbent ($x_{CaSO4,low}$ = 0.98 wt.%, $x_{CaSO4,high}$ = 1.75 wt.%). In terms of quantitative agreement, there is a difference of nearly a factor of three between the process model derived and the experimental data. It is likely that the numbers derived from TGA analysis tend towards an underestimation of the maximum average CO_2 carrying capacity, as the conditions in a TGA differ in comparison to a fluidized bed. This implies that there is no particle mixing in the TGA. Furthermore, heating rates during carbonation-calcination cycling are practically limited, which increases the exposure time of the sorbent during the calcination step. This additionally influences sorbent deactivation mechanism. Another explanation is associated with the calcination conditions in the 1 MW$_{th}$ calciner. It might be further the case that sorbent activity is reduced as a results of the prevailing calciner process conditions. As the process model does not take into account the effects of calciner operation, the reduced sorbent activity is only shows in the TGA data set.

The carbonator active space time (see Chapter 3.9.5) represents one crucial metric for the assessment of the CO_2 absorption, as it covers most of the relevant parameter of influence. Figure 4-4 shows a comparison between the results of the process model and the experimental data in terms of the nominalized CO_2 absorption efficiency of the carbonator as a function of the carbonator active space, respectively. The experimental data points were calculated by the simplification that the average molar sorbent carbonation level represents the maximum average CO_2 carrying capacity of the sorbent, as the residence time of the particles is sufficiently high to be fully converted in the fast reaction regime. The carbonator active space time derived by the process model does not make this simplification and applies the maximum average CO_2 carrying capacity instead.

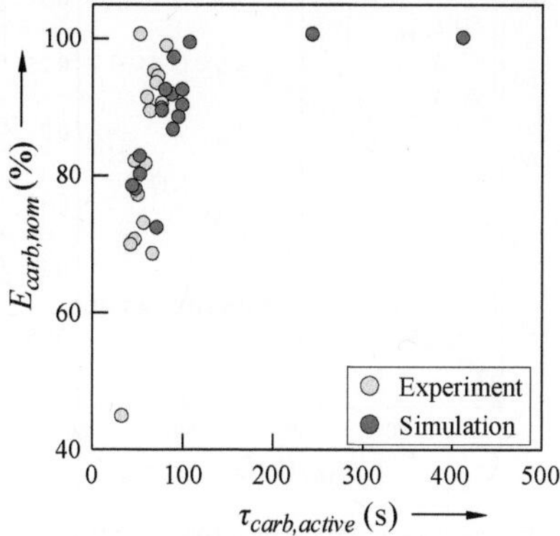

Figure 4-4: Comparison of the influence of the carbonator active space time ($\tau_{carb,active}$) on the nominalized carbonator CO_2 absorption efficiency ($E_{carb,nom}$).

Generally, the carbonator active space time derived through the process model tends towards higher values. This is due to the fact that a higher fraction of the solid population in the carbonator is regarded as active. As already shown, the process model tends to an underestimation of sorbent deactivation (see Figure 4-3). The strong decrease of the nominalized carbonator CO_2 absorption efficiency at carbonator active space times below 100 s is pronounced for both data sets.

4.3 Modell Sensitivity

A sensitivity study is carried out in order to show the influence of crucial process parameters on the CO_2 absorption efficiency of the carbonator. This analysis includes parameters that are directly controllable such as the make-up limestone feeding rate or the sorbent circulation rate. Additionally, the effect of the sulfation level of the sorbent, which is only indirectly controllable, is assessed. The boundary conditions for the sensitivity study are derived from validation point XI, which shows the best agreement between the experimental data and the process model. Table 4-3 summarizes the boundary conditions for the sensitivity analysis. A more comprehensive summary of this data point is given in Table A-3 in the appendix.

Table 4-3: Boundary conditions for the sensitivity analysis.

Parameter	F_0	F_R	F_{CO2}	$W_{s,carb}$	T_{carb}	$y_{CO2,carb,in}$
Unit	mol/s	mol/s	mol/s	kg/m²	°C	vol.%
Value	0.10	10.3	0.88	485	668	9.89

4.3.1 Make-up Limestone Feeding Rate

Figure 4-5 shows the influence of the make-up limestone mass flow ($\dot{m}_{MU}$) on the CO_2 absorption efficiency (E_{carb}) and on the maximum ($X_{max,ave}$) and average (X_{ave}) molar sorbent conversion. As the molar CO_2 load of the flue gas is kept constant, the specific make-up rate (Λ) is directly linked to the values for the make-up limestone mass flow. During the validation point, approximately 37 kg/h make-up limestone (Λ: 0.118) yields to a CO_2 absorption rate of 74.9 %. The average carbonation level (f_{carb}) is 0.623.

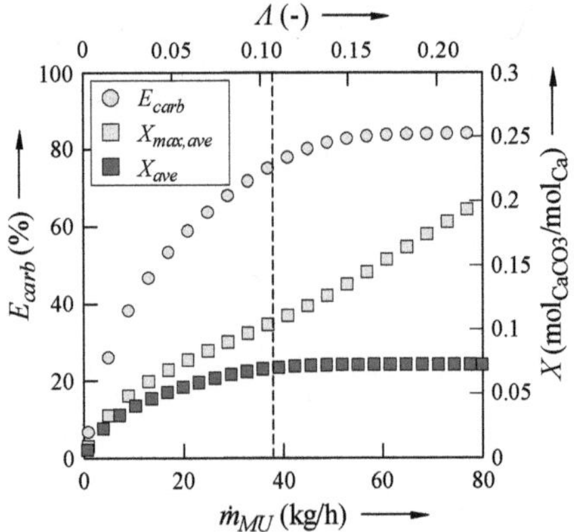

Figure 4-5: Influence of the make-up limestone mass flow ($\dot{m}_{MU}$) on the CO_2 absorption efficiency (E_{carb}) and on the maximum average ($X_{max,ave}$) and average molar sorbent conversion (X_{ave}).

The make-up limestone feeding rate influences the average reaction age (i.e. the number of complete carbonation-calcination cycles) of the sorbent (see Eq. 4-12). Accordingly, the sorbent performance is positively affected, if the make-up limestone feeding rate is raised. Furthermore, the extent of impurities (e.g. ash, $CaSO_4$) in the circulating sorbent is reduced as more material is purged to maintain the solid looping system inventory. On the other hand, the energetic process performance is deteriorated as the non-recoverable share of heat consumed by the fist calcination of the limestone increases. Once the make-up limestone feeding rate reaches approximately 50 kg/h (Λ: 0.14), the CO_2 absorption efficiency is close to its theoretical maximum at the chemical equilibrium. Thus the molar sorbent conversion remains constant at approximately 0.070. Accordingly, the average carbonation level in the carbonator decreases considerably at higher make-up limestone feeding rates.

4.3.2 Specific Sorbent Circulation Rate

Figure 4-6 shows the influence of the specific sorbent circulation rate (Φ) on the CO_2 absorption efficiency (E_{carb}), the maximum ($X_{max,ave}$) and on the average molar sorbent conversion (X_{ave}). As the molar CO_2 load of the flue gas is kept constant, the specific sorbent circulation rate is directly linked to the numbers for the mass flow rate of CaO ($\dot{m}_{CaO}$). During the validation point, a CaO mass flow rate of approximately 2,077 kg/h (Φ: 11.8) yields a CO_2 absorption rate of 74.9 %. The average carbonation level (f_{carb}) is 0.623.

The sorbent circulation rate directly alters the residence time of the particles in the carbonator and the average carbonation level. As a consequence, the maximum average sorbent conversion level increases along with higher sorbent circulation rates, since the particles being less carbonated per carbonation-calcination cycle. This effect is partly compensated by the fact, that sorbent deactivation proceeds per cycle, even with a low degree of carbonation. A sorbent circulation rate of approximately 2,500 kg/h (Φ: 14.4) would be an ideal compromise between the CO_2 absorption efficiency of the carbonator and thermal requirement of the calciner induced by the circulating sorbent. At a sorbent circulation rate of 4,000 kg/h (Φ: 22.9), only 27.5 % of the active fraction is carbonated. At lower sorbent circulation rates, the average carbonation level increases up to 82.4 %. However, even this relatively high average carbonation level is not sufficient to achieve a reasonable CO_2 absorption efficiency.

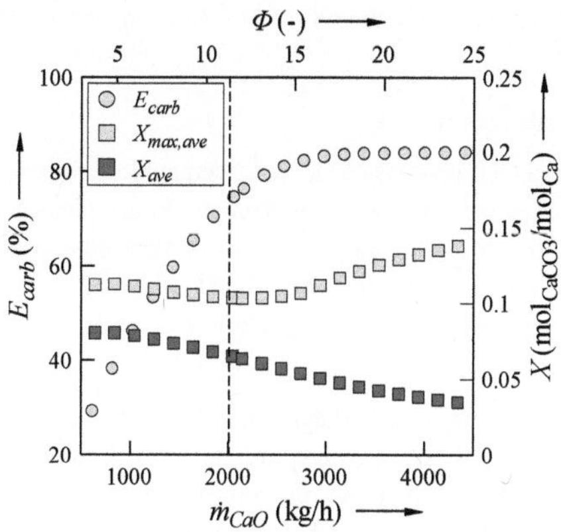

Figure 4-6: Influence of the sorbent circulation rate (Φ) on the CO_2 absorption efficiency (E_{carb}), on the maximum average (X_{max}) and average molar sorbent conversion (X_{ave}).

4.3.3 Sorbent Sulfation Level

Figure 4-7 shows the influence of the molar sorbent sulfation level (X_{sulf}) on the CO_2 absorption efficiency (E_{carb}), the maximum ($X_{max,ave}$) and on the average molar sorbent conversion (X_{ave}).

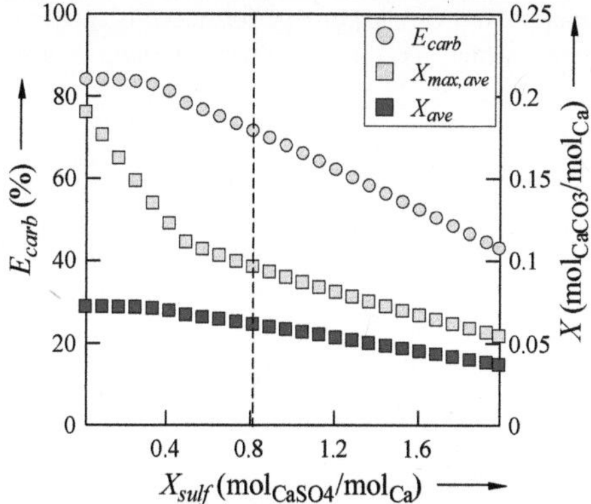

Figure 4-7: Influence of the molar sorbent sulfation level (X_{sulf}) on the CO_2 absorption efficiency (E_{carb}), on the maximum (X_{max}) and average molar sorbent conversion (X_{ave}).

During the validation point XI, the molar sulfation level is approximately 0.8, which corresponds to a CO_2 absorption rate of 74.9 %. The sorbent deactivation constant and the residual conversion capacity are directly linked to the molar sulfation level of the sorbent (see Eqs. 4-8 to 4-10). According to Grasa et al. [181], further sulfur induced sorbent deactivation is less pronounced above a critical molar sulfation level of 0.5 mol_{CaSO4}/mol_{Ca}. The implementation of this correlation can be graphically derived from the above figure. The carbonator CO_2 absorption efficiency is close to the theoretical maximum at a molar sulfation level of 0.33 mol_{CaSO4}/mol_{Ca}. If the molar sulfation level increases up to 1.98 mol_{CaSO4}/mol_{Ca}, the carbonator CO_2 absorption efficiency drops down to 42.9 %. One approach to compensate for the accumulation of $CaSO_4$ is to increase the make-up limestone feeding rate.

4.4   Process Modelling Summary

In the course of this chapter, a CaL process model has been applied for the prediction of CaL process performance. The model was initially proposed by Romano [180]. Until now, it had been considered to be the most accurate approach for the calculation of sorbent deactivation, as the sulfur induced deactivation mechanism is implemented on a high level of detail. It has been already applied in another study for the calculation of CaL process performance at the 1 MW_{th} pilot plant using fossil fuels in the calciner [241].

As the first step, the model has been modified according to the needs to fit in the experimental data. Thereafter, the model validation has been carried out in terms of a comparison of the CO_2 absorption efficiency derived by experimental data and by the process model. It was found that the majority of validation points were calculated by a relative deviation of +/- 10 %. The process model tends towards an overestimation of process performance, which might be related to the fact that the sorbent deactivation proceeds faster under the conditions of the 1 MW_{th} pilot plant.

Following the validation of the process model, a parameter study was carried out in order to show the sensitivity of the model to varying CaL process conditions in the boundary conditions of the 1 MW_{th} pilot plant. Here, the positive effect of a higher make-up limestone feeding rate and a higher sorbent circulation rate was proven. For large-scale applications of the CaL process, these parameters need to be chosen wisely, as the improved CO_2 capture performance is counteracted by a higher heat and oxygen demand in the calciner, which is results in a lower energetic process performance.

5 Techno-Economic Investigations

The retrofit of an SRF-fired CaL process onto a generic 60 MW$_{th}$ WtE plant fueled by MSW is assessed as part of this chapter. Based on a detailed thermodynamic process modeling, techno-economic key performance indicators are calculated by means of a bottom-up approach. For the appropriate modelling of the sorbent deactivation in the CaL process, the process model validated in the course of Chapter 4 is applied here. In addition to the assessment of the base case scenario, the effects of relevant technical and economical boundary conditions are addressed by means of sensitivity analyses. Furthermore, advanced heat integration concepts between CaL process and WtE plant water-steam cycle that allow for higher thermal efficiency are proposed and assessed.

5.1 Technical Evaluation Methodology

This section describes the technical evaluation methodology that is applied for the calculation of the heat and mass flows of the retrofitted WtE plant. The thermodynamic process modelling is performed in the ASPEN PLUS^{TM} 8.8 flowsheet simulation environment, using the Peng Robinson Equation of state [242]. The thermodynamic modelling includes the water-steam cycle of the reference WtE plant and the comprehensive CaL system (i.e. solid looping system, CaL water-steam cycle).

5.1.1 Reference Waste-to-Energy Plant

A generic WtE plant serves as a reference plant for the techno-economic assessment. It consists of a moving grate firing system for MSW with a nominal thermal duty of 60 MW$_{th}$. The flue gas cleaning is realized by a dry sorbent injection (DSI) unit. Accordingly, the absorption agent (e.g. sodium hydroxide, hydrated lime or limestone) is introduced to the flue gas duct as dry powder upstream of the bag filter / electrostatic precipitator [33]. The composition of the MSW (LHV: 10 MJ/kg) is summarized in Table 5-1. A biogenic carbon fraction of 65 % in MSW is assumed here [243].

Table 5-1: Composition of the municipal solid waste [82].

Parameter	Unit	Value
C	wt.%	28.9
H	wt.%	3.20
N	wt.%	0.50
S	wt.%	0.10
Cl	wt.%	0.40
O	wt.%	23.1
H$_2$O	wt.%	25.0
Ash	wt.%	18.8

The composition of the WtE plant flue gas is derived by a combustion calculation assuming an oxygen-to-fuel ratio of approximately 1.5 for the framework of the WtE plant boiler. In the subsequent heat recovery unit and the DSI system, the flue gas stream is further diluted by leakage air. The final composition of the WtE flue gas at stack conditions is summarized in Table 5-2. The flue gas is introduced to the carbonator ID-fan at approximately 135 °C and atmospheric pressure.

Table 5-2: Composition of the WtE plant flue gas at stack.

Parameter	Unit	Value
CO_2	vol.%	10.0
N_2	vol.%	70.3
H_2O	vol.%	11.4
O_2	vol.%	8.20
SO_2	ppm_v	19.0
HCl	ppm_v	6.80

Heat is recovered from the hot combustion flue gas by means of superheated steam ($\dot{m}_{steam}$: 20.2 kg/s), from which the energy is further converted into electricity via the steam turbine and the generator. The heat recovery system of the WtE plant delivers superheated steam at 40 bar and 400 °C to the inlet of the steam turbine. Figure 5-1 shows the thermodynamic layout of the WtE plant water-steam cycle. The steam turbine is equipped with three steam bleedings. One steam extraction for the preheating of combustion air (CA) and two for condensate preheating.

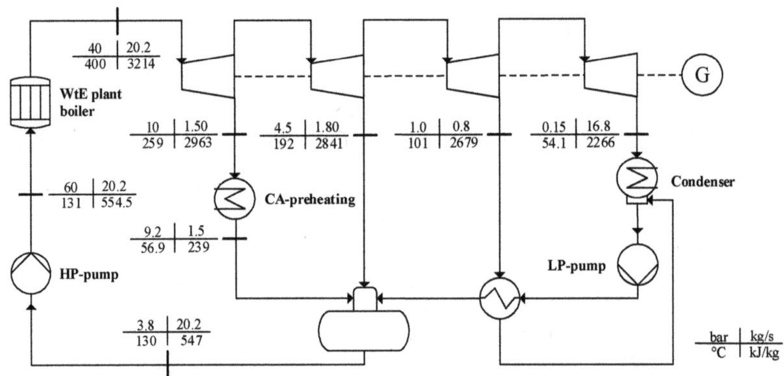

Figure 5-1: Simplified thermodynamic layout of the WtE plant water-steam cycle.

Table 5-3 summarizes the key thermodynamic characteristics of the reference WtE plant. In the framework of this study, only electricity is produced by the WtE plant, the delivery of heat is not taken into account.

Table 5-3: Thermodynamic characteristics of the reference WtE plant.

Parameter	Unit	Value
MSW mass flow, $\dot{m}_{MSW}$	kg/s	6.00
Gross power output, $P_{el,gross}$	MW_e	14.4
Net power output, $P_{el,net}$	MW_e	12.3
Total CO_2 emissions, $e_{CO2,tot}$	g_{CO2}/kWh$_e$	1680
Fossil CO_2 emissions, $e_{CO2,foss}$	g_{CO2}/kWh$_e$	590
Gross electrical efficiency, η_{gross}	%	26.2
Net electrical efficiency, η_{net}	%	22.8

5.1.2 Carbonate Looping Solid Looping System

A flowsheet of the CaL solid looping system is shown in Figure 5-2. The main components are the carbonator and the calciner. The flue gas of the WtE plant is forced through the carbonator by an ID-fan. After solid and gas phase separation in the carbonator cyclone, sensible heat of the CO_2-depleted flue gas is transferred to the CaL water-steam cycle (Carb II). Subsequently, fly ash is removed before the CO_2-depleted flue gas stream is released to the environment. In order to keep the operation temperature of the carbonator on the desired level, heat is removed by the evaporation of feedwater, either in water-walls directly attached to the carbonator riser or in an external fluidized bed heat exchanger (Carb I).

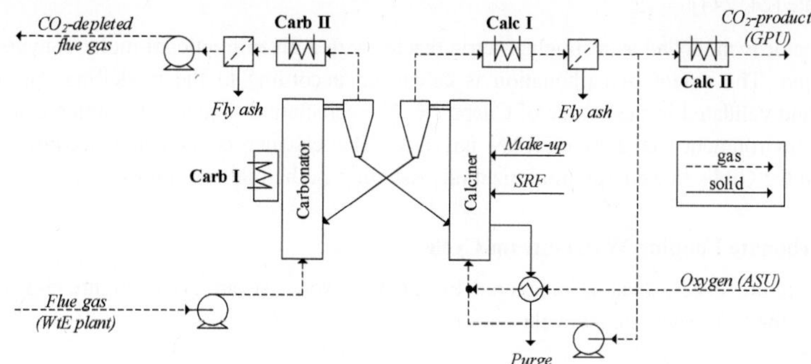

Figure 5-2: Flowsheet of the CaL solid looping system.

The required heat in the calciner is provided by oxyfuel combustion of SRF. As for oxyfuel combustion systems, off-gas is recirculated and mixed with technically pure oxygen before this gas stream is fed to the calciner as fluidization and oxidation agent. Similarly to the carbonator, the gas/solid stream is routed through a cyclone separator. The regenerated solid stream is returned to the carbonator. The temperature of the calciner off-gas is reduced in a first heat exchanger (Calc I) followed by the separation of fly ash. Thereafter, part of the off-gas is recirculated to the inlet of the calciner, whereas the other part is further cooled (Calc II) and subsequently introduced to the GPU. The make-up limestone and the SRF are directly fed to the calciner, whereas the technically pure oxygen is preheated by the spent material purged from the solid loop.

Table 5-4 summarizes the boundary conditions for the modelling of the CaL solid looping system.

Table 5-4: Boundary conditions for process modelling of the CaL solid looping system.

System	Parameter	Unit	Value
Carbonator	Temperature, T_{carb}	°C	650
	Specific solid inventory, $W_{s,carb}$	kg/m²	1000
	CO_2 absorption rate, E_{carb}	%	80
	Outlet temperature of CO_2 depleted flue gas, $T_{FG,out}$	°C	120
	Superficial gas velocity, $u_{0,carb}$	m/s	5.0
	Pressure drop, Δp_{carb}	mbar	220
	Specific sorbent circulation, Φ	-	10
Calciner	Temperature, T_{calc}	°C	900
	Temperature recirculation gas, T_{reci}	°C	200
	Outlet temperature flue gas, $T_{CO2,out}$	°C	120
	Superficial gas velocity, $u_{0,calc}$	m/s	5.0
	Pressure drop, Δp_{calc}	mbar	160
	Oxygen concentration (inlet), $y_{O2,calc,in}$	vol.%	60
	Oxygen concentration (outlet), $y_{O2,calc,out}$	vol.%	3.5
	Fuel ash losses	%	30
ASU	Specific power consumption, $P_{el,AUS}$	kWh$_e$/kg$_{O2}$	220
	Oxygen purity, y_{O2}	vol.%	95
GPU	Specific power consumption, $P_{el,GPU}$	kJ$_e$/kg$_{CO2}$	395
	CO_2 recovery rate, R_{CO2}	%	97

During the course of process modelling, the specific sorbent circulation rate (Φ) is kept constant, while the specific make-up rate (Λ) is varied iteratively in order to achieve the desired CO_2 absorption efficiency in the carbonator. The technical characteristics of the ASU and the GPU are accounted for by specific power consumption and each product characteristics, such as oxygen purity and CO_2 recovery rate [24, 244].

The carbonator is modelled as a stoichiometric reactor with a given fractional molar conversion for each reaction. The extent of carbonation is calculated according to the modelling approach as described and validated in the course of Chapter 4. The calculation sequence is implemented in the flowsheet environment via a FORTRAN interface. The calciner is modelled according to the approach of the Gibbs free energy minimization, assuming isothermal conditions.

5.1.3   Carbonate Looping Water-Steam Cycle

High temperature excess heat is recovered by the CaL water-steam cycle. Figure 5-3 shows a flowsheet of the CaL water-steam cycle.

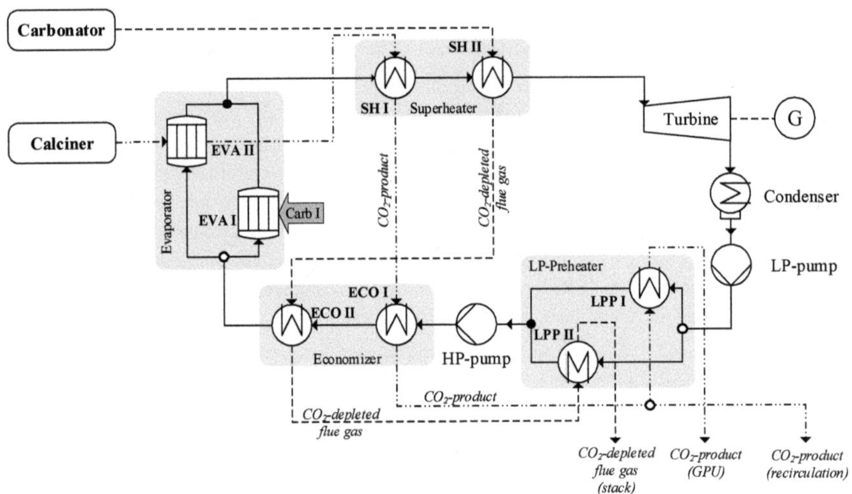

Figure 5-3: Flowsheet of the CaL water-steam cycle.

The largest share of the sensible heat of the CO_2-depleted flue gas at the outlet of the carbonator at 650 °C and the CO_2-rich flue gas at the outlet of the calciner at 900 °C are transferred to the CaL water-steam cycle. In order to minimize the risk of chlorine-induced high-temperature corrosion at the heat exchanger surfaces contained in the calciner convective pass, the CO_2-product at the outlet of the calciner is first cooled by means of evaporator tubes (EVA II) before it is further cooled in the first superheater (SH I) and in the first economizer (ECO I). The carbonator convective pass consists of the secondary superheater (SH II) and the secondary economizer (ECO II). Part of the preheated condensate is fed to the first evaporator (EVA I), that is directly attached at the carbonator walls, in order to keep its temperature at the desired level. The boundary conditions for process modelling of the CaL water steam-cycle are summarized in Table 5-5.

Table 5-5: Boundary conditions for process modelling of the CaL water-steam cycle.

Parameter	Unit	Value
Fan isentropic efficiency	%	85
Fan mechanical efficiency	%	95
Steam turbine isentropic efficiency	%	78
Steam turbine mechanical efficiency	%	97
Outlet temperature CO_2-depleted flue gas	°C	120
Temperature recirculation gas	°C	200
Outlet temperature CO_2-product gas	°C	120
Live steam temperature	°C	485
Live steam pressure at steam turbine inlet	bar	60
Pressure drop superheater, incl. turbine inlet valve	%	10
Pump efficiency	%	80
Gas phase pressure drop convective pass	mbar	35
Evaporator pressure drop	bar	20
Condenser pressure	mbar	0.07

The steam turbine isentropic efficiency is adapted according to Consonni et al. [245], who investigated turbine efficiencies in dependence of gross electric power output. Especially in small-scale steam turbines, the isentropic efficiency is lower than in large-scale power generation applications. The remaining parameters were chosen according to relevant guidelines for the evaluation of thermodynamic CCS studies [22, 244].

5.2 Economical Evaluation Methodology

This section describes the economical evaluation methodology that is considered for the calculation of the costs related to the CO_2 capture and conditioning process. Costs related to the transportation and storage of CO_2 are not taken into account here. Particular focus is on the determination of the costs for atmospheric CO_2 removal while considering the organic waste fractions in MSW and SRF. The economic evaluation is based on the calculation of costs for the supply of electricity as primary output product of the WtE plant. The total costs consists of the capital expenditures (CAPEX) expressed as total plant costs (TPC) and the operating expenditures (OPEX). This work does not perform economic analysis of the WtE plant without CO_2 capture. Therefore, the levelized cost of electricity of 80 EUR/MWh$_e$ is assumed as a constant [246, 247].

5.2.1 Capital Expenditures

The capital expenditures (CAPEX) are calculated by a bottom-up approach, which takes into account the investment cost for each component. Figure 5-4 shows the applied methodology for the calculation of the total plant costs (TPC).

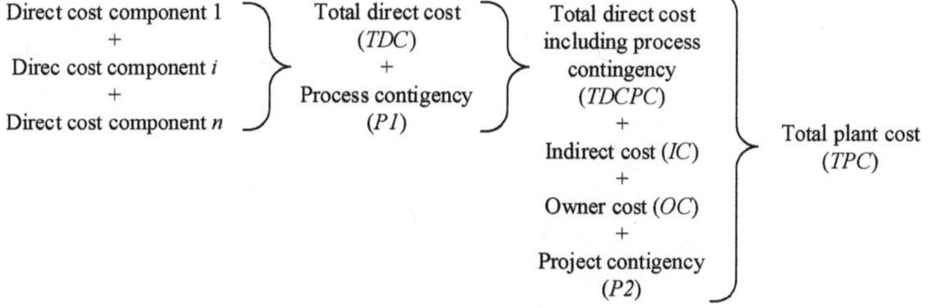

Figure 5-4: Methodology for the calculation of the total plant costs, adopted from [24].

The direct costs of standard components are derived from the ASPEN PLUS^{TM} Economic Analyzer, while costs of non-standard components are calculated by the specification according to Table A-4 in the appendix.

Subsequently, process contingencies (P1) are added to the total direct costs to calculate the total direct cost including process contingency (TDCPC). Finally, indirect costs (IC, 14 % of TDCPC), owner costs (OC, 7 % of TDCPC) and project contingencies (P2) were added to the TDCPC in order to determine the total plant cost (TPC).

Table 5-6 summarizes the process contingency and project contingency levels for each technology. The process contingency level is related to the maturity of each technology. The CaL process thus shows a relatively high process contingency level (P1$_{CaL}$: 32 %) in contrast to the equipment related to the water-steam cycle or the auxiliary components (e.g. fuel and limestone feeding systems).

Table 5-6: Process and project contingency levels [214, 248].

Technology	Process contingency, P1 (% of TDC)	Project contingency, P2 (% of TCDPC)
CaL process	32	15
Water-steam cycle	5	15
Auxiliaries	5	15
GPU	15	15

5.2.2 Operational Expenditures

The operational expenditures (OPEX) consists of fixed and variable operational costs. Figure 5-5 shows the applied methodology for the calculation of the operational expenditures.

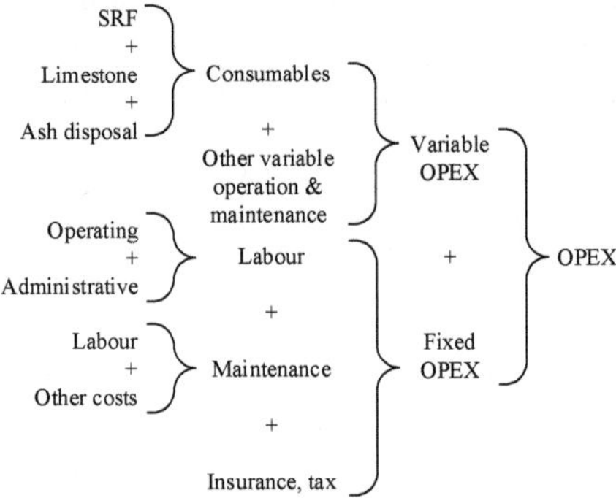

Figure 5-5: Methodology for the calculation of the operational expenditures, adopted from [24].

The fixed operational costs (FOC) cover the cost for preventive and corrective maintenance (2 % of TPC), insurance (2 % of TPC) and local taxes (0.5 % of TPC) [249]. Furthermore, the cost of labor consist of three labor classes (operating labor, maintenance labor and administrative labor). The cost for operating labor is estimated based on the expected number of employees needed to run the CO_2 capture process and a fully burdened annual cost of 60.000 EUR/a per employee [250]. The maintenance labor is assumed to correspond to 40 % of the total maintenance annual cost. The administrative labor is assumed to be 30 % of the operating and maintenance labor cost.

The variable operating costs (VOC) of the CO_2 capture process are dependent on the process performance and include the use of raw materials such as limestone and SRF to deliver the desired product. The VOC are evaluated based on utility consumption obtained from the process simulation (i.e. mass flows) and the corresponding utility costs as shown in Table 5-7.

Table 5-7: Cost of utilities.

Utility	Unit	Cost
SRF	EUR/t	-25
Limestone	EUR/t	10
Ash disposal	EUR/t	0

While most of the utilities are commonly used in techno-economic CCS studies, it is worth noting that the price for SRF varies greatly in literature. This number depends on the regulatory framework, gate fees and the availability of suitable waste streams. Exemplarily, a study for the UK showed a revenue per mass of SRF utilized in the range of approximately 30 - 105 EUR/t$_{SRF}$ [21]. For the base case scenario, a SRF price of -25 EUR/t is assumed, which seems to be reasonable, as the SRF shows a relatively high quality class according to the EU specification (see Chapter 2.1.2).

5.2.3   Key Performance Indicators

In this chapter, the thermodynamic and economic key performance indicators (KPI) relevant for process evaluation are introduced. Furthermore, the hypothetical reference state that takes into account the indirect electricity supply decarbonization by the additional supply of low-carbon power by the CaL process is introduced.

5.2.3.1   Definition of the Hypothetical Reference State

One distinguished characteristic of the CaL process is the supply of additional low-carbon power once this process is retrofitted to an existing power or industrial plant. Therefore, the KPIs of the corresponding reference state (i.e. the host plant without CaL process) need to be extended in order to account for the indirect decarbonization. The net electrical efficiency in the hypothetical reference state, $\eta_{net,ref}$, is calculated as follows:

$$\eta_{net,ref} = (1 - \delta_{CaL}) * \eta_{net,WtE} + \delta_{CaL} * \eta_{net,SRF} \qquad (5\text{-}1)$$

Here, δ_{CaL} is the share of the additional net power supplied to the grid calculated according to Eq. 5-2, and $\eta_{net,SRF}$ is the reference net electrical efficiency for SRF fueled power plants based on the currently available state-of-the-art technology (see Table 5-8), and $\eta_{net,WtE}$ is the net electrical efficiency of the reference WtE plant without CCS. This approach differs from the one described in Figure 2-18 and Eq. 2-38, as two different types of fuel are utilized in the host unit and in the CaL calciner.

$$\delta_{CaL} = \frac{\Delta P_{el,net,CaL}}{P_{el,net,ref} + \Delta P_{el,net,CaL}} \qquad (5\text{-}2)$$

Here, $\Delta P_{el,net,CaL}$, is the increase in total plant net power after the retrofit of the CaL process.

Similarly, the specific fossil CO_2 emissions in the hypothetical reference state, $e_{CO2,foss,ref}$, are calculated as follows:

$$e_{CO2,foss,ref} = (1 - \delta_{CaL}) * e_{CO2,foss,WtE} + \delta_{CaL} * e_{CO2,foss,SRF} \qquad (5\text{-}3)$$

Here, $e_{CO2,foss,SRF}$ are the specific fossil CO_2 emissions for SRF fueled power plants based on the currently available state-of-the-art technology (see Table 5-8) and $e_{CO2,foss,WtE}$ are the specific fossil CO_2 emissions of the reference WtE plant without CCS.

In accordance to the aforementioned KPIs, the LCOE$_{ref}$ are calculated with respect to additional power that is supplied by the CaL unit:

$$LCOE_{ref} = (1 - \delta_{CaL}) * LCOE_{WtE} + \delta_{CaL} * LCOE_{SRF} \tag{5-4}$$

Here, LCOE$_{SRF}$ are the LCOE for SRF fueled power plants based on the currently available state-of-the-art technology (see Table 5-8), and LCOE$_{WtE}$ are the LCOE of the reference WtE plant without CCS.

Table 5-8: Techno-economic boundary conditions for state-of-the-art power plants [55].

Parameter	Coal	Natural gas	SRF	WtE
Net electrical efficiency, %	45	61	25	22.8
Specific fossil CO_2 emissions, g_{CO2},kWh$_e$	763	325	771	590
Specific biogenic CO_2 emissions, g_{CO2},kWh$_e$	-	-	514	1090
Levelized cost of electricity, EUR/MWh$_e$	51.6	52.9	80.0	80.0

In order to show the effects of the indirect decarbonization by the additional low-carbon power of the CaL process, the development of LCOE$_{ref}$ (Graph a), $\eta_{net,ref}$ (Graph b) and $e_{CO2,foss,ref}$ (Graph b) as a function of the additional net power ratio (δ) are presented by Figure 5-6. As the state-of-the-art parameters for SRF and MSW fuelled power plants are rather similar, this assessment also covers the cases of coal and natural gas as CaL supplementary fuels. Table 5-8 summarizes the applied techno-economic boundary conditions. For typical CaL process applications in the power sector, the additional net power ratio ranges from 29 - 41 % [198, 202, 206].

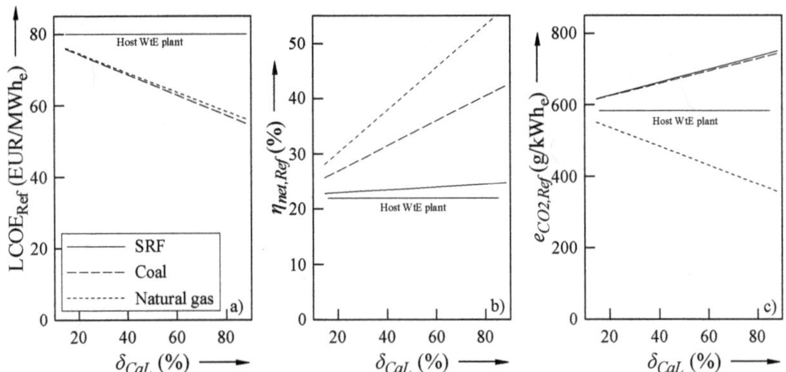

Figure 5-6: Influence of additional low-carbon net power on the LCOE (a), net electrical efficiency (b) and on the specific fossil CO_2 emissions (c) as a function of the additional net power ratio (δ_{CaL}).

From Figure 5-6, it is clear that the consideration of the hypothetical reference state is significant for the techno-economic CaL process evaluation. For the case of SRF, the net electrical efficiency and the related specific fossil CO_2 emissions increase with a higher additional net power ratio. This implies that the net electrical efficiency drop and the SPECCA of the retrofitted WtE plant are underestimated in case of neglecting the hypothetical reference state. Due to the fact that the relevant fuel properties of MSW and SRF are relatively similar to each other, the hypothetical reference state is relatively close to the numbers of the WtE plant without CCS. However, in case of coal or natural gas as supplementary fuel, the differences are more distinct. Assuming, for example, that natural gas is utilized as supplementary fuel and that the CaL process supplies 40 % of the net power, the hypothetical net electrical efficiency is increased by approximately 15 %-points. In order to classify these figures, it needs to be noted that the net electrical efficiency drop associated to a CaL process retrofit onto a state-of-the-art coal-fired power plant ranges only between 4 to 7 %-points.

5.2.3.2 Thermodynamic KPI

Due to the dedicated water-steam cycle of the CaL process, additional power is supplied by the utilization of CaL process excess heat. In order to quantify the additional gross power output, the CaL gross power ratio, PR_{CaL}, is introduced according to Eq. 5-5. Doing so, the CaL water-steam cycle performance is directly assessable without the effects of auxiliary power consumption.

$$PR_{CaL} = \frac{P_{el,CaL,gross}}{P_{el,WtE,gross} + P_{el,CaL,gross}} \tag{5-5}$$

Here, $P_{el,gross,CaL}$ is the gross electrical power output of the CaL water-steam cycle and, $P_{el,gross,WtE}$ is the gross electrical power output of the reference WtE plant.

One important metric for the evaluation of CCS applications is the specific energy consumption per CO_2 avoided (SPECCA). This metric weights the efficiency drop for the supply of a service (e.g. heat or power) by the amount of the CO_2 that is avoided in the course of the supply for this service. For the application in power plants for the delivery of electricity only, it is calculated as follows:

$$SPECCA = \frac{\dfrac{1}{\eta_{net,CaL}} - \dfrac{1}{\eta_{net,ref}}}{e_{CO2,foss,ref} - e_{CO2,CaL}} \tag{5-6}$$

Here, $\eta_{net,CaL}$ is the net electrical efficiency of the WtE plant with CaL process and, $\eta_{net,ref}$ is the net electrical efficiency in the hypothetical reference state. Furthermore, $e_{CO2,foss,ref}$ are the specific fossil CO_2 emissions in the hypothetical reference state and, $e_{CO2,CaL}$ states the specific CO_2 emissions of the WtE plant retrofitted by the CaL process, calculated by Eq. 5-7:

$$e_{CO2,CaL} = e_{CO2,foss,CaL} - e_{CO2,capt,bio} \tag{5-7}$$

Here, $e_{CO2,foss,CaL}$ are the specific fossil CO_2 emissions of the retrofitted system and $e_{CO2,capt,bio}$ is the specific amount of captured biogenic CO_2 in the CaL retrofit case, respectively. The specific amount of captured biogenic CO_2 is accounted for negatively due to the CO_2 uptake during the growth of biogenic feedstock. Thus, CO_2 is removed from the atmosphere which results is net negative CO_2 emissions.

5.2.3.3 Economic KPI

The levelized cost of electricity (LCOE) measures the unit cost of a power plant with and without CO_2 capture, and approximates the average discounted electricity price over the project duration that would be required as income to match the net present value of costs of the project. The LCOE for the retrofitted WtE plant, $LCOE_{CaL}$ (EUR/MWh$_e$) are calculated according to Eq. 5-8 [204, 251].

$$LCOE_{CaL} = \frac{TPC * FCF + FOC}{CF * 8766 + MW} + VOC + HR * FC \tag{5-8}$$

Here, FCF is the fixed charge factor (Eq. 5-9), CF is the capacity factor ($CF = 0.85$), MW is the net power of the unit and HR is the net power plant heat rate. The TPC are derived according to Figure 5-4. The calculation of the fixed operational expenditure (FOC) and of the variable operational expenditures (VOC) is described elsewhere (see Chapter 5.2.1 and 5.2.2).

$$FCF = \frac{r(1+r)^t}{(1+r)^t - 1} \qquad\qquad (5\text{-}9)$$

Here, r is the interest rate ($r = 0.08$) and, t is the economic lifetime ($t = 25$ a) of the project.

The cost of CO_2 avoided (CAC) which is obtained by comparing the LCOE and specific CO_2 emission rate to the atmosphere of the WtE plant with and without CaL process are calculated according as follows:

$$CAC = \frac{LCOE_{CaL} - LCOE_{ref}}{e_{CO2,ref} - e_{CO2,CaL}} \qquad\qquad (5\text{-}10)$$

5.3 Techno-Economic Process Assessment

This chapter provides a techno-economic assessment of the CaL process retrofit on the reference WtE plant. The thermodynamic assessment focusses on the thermal and electrical system characteristics, while considering all relevant subsystems of the CaL process (e.g. solid looping cycle, water-steam cycle, ASU, GPU). Furthermore, a sensitivity analysis shows the effect of CaL process conditions and live steam parameters of the CaL water-steam cycle on the overall process performance. In addition to that, a novel concept for CaL process excess heat recovery by means of external superheating of the WtE plant power cycle is proposed and assessed.

The economical process assessment discusses the LCOE and the CAC for the retrofitted WtE plant. A breakdown of LCOE and the corresponding CAC is presented. Uncertainties in the boundary conditions related to the early stage of development of the CaL process are accounted for in the course of the evaluation of economical extreme case scenarios. Additionally, the economics of the system are discussed, while the supply of negative CO_2 emissions is considered as an additional product of the CaL retrofitted WtE plant in the future. Consequently, the comparison is extended in order to take into account competing technologies for atmospheric CO_2 removal (i.e. negative CO_2 emission technologies, see Chapter 2.2.4).

5.3.1 Thermodynamic Assessment

In the course of this Chapter, a thermodynamic assessment of the full-scale process is carried out. This covers the evaluation of the base case scenario and a sensitivity analysis of crucial boundary conditions made in the course of process modelling. Furthermore, the effects of the CaL water-steam cycle parameters are addressed and advanced heat integration concepts are proposed and evaluated.

5.3.1.1 Thermodynamic Base Case Assessment

Table 5-9 summarizes the thermodynamic performance of the WtE plant retrofitted by the CaL process. The thermal power input and the electrical power output increase more than twofold in the case of a CaL retrofit. Due to the higher live steam parameters of the CaL water-steam cycle, the gross electrical efficiency increases by 0.4 %-points. However, the net electrical efficiency is significantly reduced due to the additional auxiliary power requirement mainly associated to the ASU and the GPU. As both types of fuel that are considered in this study contain a significant amount of organic waste fractions, the extent of biogenic CO_2 in the CO_2 product stream leads to net negative CO_2 emissions. It needs to be noted that the net electrical efficiency is still relatively high in comparison with other technologies so far considered for CO_2 capture in the framework of WtE plants, such as MEA-based CO_2 capture (η_{net} = 7.3 %), advanced solvent-based CO_2 capture (η_{net} = 9.2 %) or membrane-based CO_2 capture (η_{net} = 3.7 %) [252].

Table 5-9: Thermodynamic performance of the WtE plant retroffited by the CaL process.

Parameter	Unit	WtE	WtE + CaL
Thermal power, P_{th}	MW_{th}	60.0	132
Gross electrical power, P_{el}	MW_e	15.7	35.1
Gross electrical efficiency, η_{gross}	%	26.2	26.6
Auxiliary power consumption, $P_{el,aux}$[13]	MW_e	2.10	3.73
ASU power consumption, $P_{el,ASU}$	MW_e	-	5.19
GPU power consumption, $P_{el,GPU}$	MW_e	-	5.13
Net electrical power, $P_{el,net}$	MW_e	13.6	21.0
Net electrical efficiency, η_{net}	%	22.7	16.0
CO_2 capture rate, E_{tot}	%	-	91.1
Specific CO_2 emissions, e_{CO2}	g_{CO2}/kWh_e	590	-925

In order to quantify the different origin of fossil and biogenic CO_2, a summary of the specific CO_2 flows in the CaL system is given by Table 5-10.

Table 5-10: Specific CO_2 flows involved in the WtE+CaL system[14].

CO_2 quantity	Unit	Total	Fossil	Biogenic
CO_2 formation WtE plant (MSW)	g_{CO2}/kWh_e	1088	382	707
CO_2 formation CaL calciner (SRF)	g_{CO2}/kWh_e	1093	656	437
CO_2 formation CaL calciner (Make-up)	g_{CO2}/kWh_e	259	259	-
CO_2 captured	g_{CO2}/kWh_e	2222	1220	1003
Net CO_2 emissions	g_{CO2}/kWh_e		-925	
CO_2 avoidance	g_{CO2}/kWh_e		1576	

5.3.1.2 Effect of CaL Process Conditions

The CaL process conditions (e.g. sorbent deactivation, oxyfuel combustion conditions in the calciner) have a significant influence on the heat demand of the calciner and subsequently on the overall thermodynamic process characteristics. The heat demand is mainly determined due to the heating of the incoming material streams and due to the heat of reaction for the calcination of limestone. Table 5-11 shows the boundary conditions for the assessment of the influence of CaL process conditions.

Table 5-11: Boundary conditions for the assessment of the influence of CaL process conditions.

Parameter	Unit	Worst Case	Base Case	Best case
Sorbent residual conversion capacity, X_r	-	0.060	0.075	0.090
Sorbent decay constant, k_r	-	0.624	0.520	0.416
O_2-concentration calciner inlet, $y_{O2,calc,in}$	vol.%	40	60	80
O_2-concentration calciner outlet, $y_{O2,calc,out}$	vol.%	4.5	3.5	2.5
Cyclone fuel ash losses	%	20	30	40
Sorbent heat exchanger pinch temperature	K	-	-	100
Specific ASU power consumption, $P_{el,ASU}$	kWh_e/t_{O2}	240	220	200

The parameters of the sensitivity analysis were chosen in order to address the most crucial assumption made during the course of CaL process modelling. This is for instance the degree of sorbent deactivation, the oxygen concentration at the inlet and at the outlet of the calciner or the loss of fuel ash throughout the calciner cyclone. All of these parameter significantly influence the thermal requirement of the calciner, either directly, as for the oxygen concentrations, or indirectly, due to the higher make-up feed needed to compensate for the accumulation of fuel ash or forced sorbent

---

[13] Including ID-fans and feedwater pumps of the CaL system.
[14] The net power capacity of the retrofitted WtE plant is considered for the calculation of the specific CO_2 flows.

deactivation. Additionally, the transfer of sensible heat between the two major solid fluxes is applied under the best case conditions. Here, the partly carbonated sorbent stream fed to the calciner from the carbonator is preheated by calcined material leaving the calciner. It was already found that this approach significantly improves the CaL process performance [253]. Table 5-12 summarizes the thermodynamic results for the different CaL process conditions.

Table 5-12: Thermodynamic results for different CaL process conditions.

Parameter	Unit	Worst Case	Base Case	Best case
Specific make-up ratio, Λ	-	0.294	0.238	0.181
Specific circulation ratio, Φ	-	10	10	10
CO_2 absorption rate, E_{carb}	%	80	80	80
Total CO_2 capture rate, E_{tot}	%	91.9	91.1	89.8
$CaCO_3$ content at carbonator outlet, x_{CaCO3}	wt.%	10.4	10.5	10.8
$CaSO_4$ content at calciner outlet, x_{CaSO4}	wt.%	2.25	2.45	2.59
Ash content at calciner outlet, x_{Ash}	wt.%	21.3	20.1	17.9
Oxygen flow to the calciner, $\dot{m}_{O2}$	kg/s	7.83	6.55	4.94
Calciner thermal input, $P_{th,calc}$	MW_{th}	83.9	71.7	55.4
CaL process heat ratio, HR_{CaL}	-	0.583	0.545	0.480
CaL process gross power ratio, PR_{CaL}	-	0.590	0.552	0.490
ASU power demand, $P_{el,ASU}$	MW_e	6.77	5.18	3.56
GPU power demand, $P_{el,GPU}$	MW_e	5.76	5.13	4.36
Total net power output, $P_{el,net,tot}$	MW_e	21.9	21.1	19.4
Total net electrical efficiency, $\eta_{net,tot}$	%	15.2	16.0	16.8
Specific heat demand, Q_{spec}	$MJ_{th}/kg_{CO2,capt}$	5.81	5.53	4.96
Specific oxygen demand, $O_{2,spec}$	$kg_{O2}/kg_{CO2,capt}$	0.543	0.504	0.443
SPECCA	$MJ_{th}/kg_{CO2,av}$	5.20	4.56	3.94

Under best process conditions, the calciner heat duty is reduced by 16.3 MW_{th} compared to the reference state. This results in a specific heat demand of 4.96 $MJ_{th}/kg_{CO2,capt}$ and SPECCA of 3.94 $MJ_{th}/kg_{CO2,av}$. Controversially, worst process conditions lead to an increase of the thermal calciner duty by 12.2 MW_{th}, which further increases the specific heat demand (5.81 MJ_{th}/kg_{CO2}) and the SPECCA (5.20 $MJ_{th}/kg_{CO2,av}$). Along with the calciner thermal duty, the auxiliary power consumption of the GPU increases as a consequence of a higher CO_2-product mass flow that needs to be purified and compressed. The sorbent deactivation mechanism is assumed to be less distinct under improved process conditions. Accordingly the specific make-up ratio shows the lowest value in the best case (Λ_{Best} = 0.181), even though the mass fraction for $CaSO_4$ shows the highest value under best case conditions.

5.3.1.3 Effect of CaL Water-Steam Cycle Parameters

The excess heat of the CaL process is recovered by means of a dedicated water-steam cycle. In the case of large-scale CaL process applications (P_{th} > 1000 MW_{th}), the live-steam parameters (e.g. pressure, temperature) were chosen according to the current supercritical state-of-the-art technology. A highly efficient heat conversion into electricity is thus feasible. In the present thermal dimensions ($P_{th} \sim$ 50 MW_{th}), supercritical live steam parameters are neither technically applicable nor economically reasonable. Thus, subcritical live steam parameters are applied here.

As the characteristics of WtE plants vary greatly in terms of size, the potentially retrofitted CaL process and its water-steam cycle vary similarly. According to Baehr [254], the live steam parameters and water-steam cycle layout (i.e. superheater and reheater configuration) are strongly connected to the gross power output of the steam turbine. In order to quantify the effects of CaL process water-steam cycle parameters on the overall process performance, three different sets of live steam parameters according to Table 5-13 are investigated.

Table 5-13: Boundary conditions for the assessment of the CaL water-steam cycle parameter.

Parameter	Unit	Case I	Case II (Base)	Case III
Live steam temperature, T_{LS}	°C	450	485	520
Live steam pressure at turbine inlet, p_{LS}	bar	40	60	80
Turbine isentropic efficiency, η_{isentr}	%	75.0	78.0	80.8

Dependent on the expected size of the steam turbine, the corresponding isentropic efficiency is varied accordingly [245]. As the available excess heat from the CaL process is not altered throughout the investigation, the mass flow of the feedwater in the CaL water steam cycle needs to be iteratively adjusted in the process modelling step in order to match the desired outlet temperature of the CaL gas flows.

Table 5-14 summarizes the thermodynamic results of this investigation. Related to these results, Figure 5-7 shows the heat transfer curve and the heat distribution of the CaL water-steam cycle heat exchanger units for case I and case III.

Table 5-14: Thermodynamic results for different CaL water-steam cycle parameters.

Parameter	Unit	Case I	Case II (Base)	Case III
CaL gross power, $P_{el,gr,CaL}$	MW$_e$	17.7	19.4	21.0
CaL gross electrical efficiency, $\eta_{gross,CaL}$	%	24.7	27.0	29.3
CaL process gross power ratio, PR_{CaL}	-	0.530	0.552	0.572
Total net power, $P_{el,net,tot}$	MW$_e$	19.5	21.1	22.7
Total net electrical efficiency, $\eta_{net,tot}$	%	14.8	16.0	17.2
SPECCA	MJ$_{th}$/kg$_{CO2,av}$	5.46	4.56	3.75

The process characteristics are significantly influenced by the choice of the water-steam cycle parameters. An operation at higher live steam parameters (520 °C / 80 bar) raises the CaL process gross power output to 21.0 MW$_e$ ($\eta_{gross,CaL}$ = 29.3 %), which represents an increase by 18.6 % compared to the conditions of case I. Concerning the SPECCA, the improvements in Case III are approximately 17.7 % compared to the base case scenario. This is related to the fact that more power is produced per quantity of CO_2 that is emitted. In other words, less fuel needs to be supplied for the same quantity of product (e.g. electricity). On the other hand, the CaL gross power output is lowered by 11.4 % and the associated SPECCA are increased by 19.7 % for live steam parameter according to Case I.

The heat transfer curve confirms that it is thermodynamically feasible to operate the CaL water-steam cycle in the given boundary conditions as the minimum required temperature difference between hot and cold streams is given throughout the comprehensive profile. It can be seen that the temperature difference at the inlet of the second economizer (ECO 2) shows the lowest value in case of the high live steam parameters. This points shifts towards the inlet of the first superheater (SH 1) for the lowest considered live steam parameters. At the case of higher live steam parameters, the evaporation of the feedwater occurs at a higher pressure and temperature level, which lowers the corresponding specific heat demand for steam generation. At the conditions of Case I, 55.5 % of the total CaL process excess heat is used for evaporation, this number reduces to 45.1 % for Case III. Simultaneously, the share of heat transferred in the feedwater preheating section increases from 30.5 % (Case I) to 37.4 %

(Case III). The aforementioned results are also in part related to the fact the total live steam mass flows varies between 15.8 kg/s (Case III) and 23.7 kg/s (Case I).

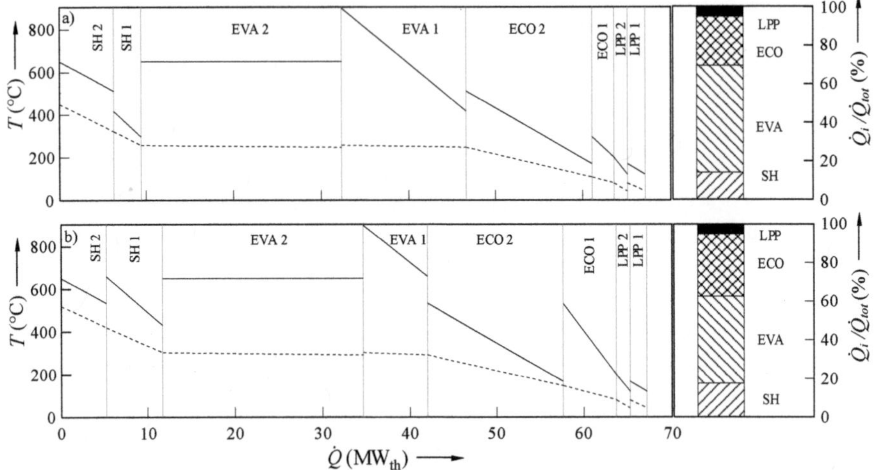

Figure 5-7: Heat transfer curve of the CaL water-steam cycle (left) and heat distribution (right), for the following live steam parameter sets: 450 °C / 40 bar (a), 520 °C / 80 bar (b).

Despite the fact that, as already shown, a large part of the fuel chlorine is absorbed by the solid phase in the calciner (see Chapter 3.11.6), there is a risk of chlorine-induced high-temperature corrosion on surfaces of the heat exchanger tubes in the heat recovery path of the calciner. The process of high temperature corrosion is a highly complex and multi-layered phenomena which is not fully understood until now [255, 256]. It is known that the temperature of the tube wall and of the surrounding gas phase have an influence on the development of corrosion [257]. The extent or the appearance of corrosion is, among others, a function of these two temperature values. In this regard, the "Flingern" diagram is often applied for corrosion characterization in the framework of WtE plants [258]. Here, the operational characteristics (e.g. wall temperature, gas phase temperature) of corrosion exposed tube walls are marked in a diagram with three distinguished corrosion regimes (e.g. heavy corrosion, transition, no corrosion).

Thus, according to the heat transfer curve, the most critical corrosion conditions prevail at the outlet of the first evaporator (EVA 1) and at the outlet of the first superheater (SH 1). Figure 5-8 shows the operational range of these heat exchangers for the process conditions of Case 1 and 3, respectively. For the calculation of the tube wall temperature, a temperature difference of 20 K is applied.

In Case I, the temperature of the steam is relatively low even at the outlet of the first superheater. The evaporator and the superheater are thus fully located in the non-corrosive regime. Once the highest live steam parameters (Case III) are applied, the corresponding outlet temperatures of superheater and evaporator increase, which leads to an operation slightly inside the transition regime.

It needs to be noted that the original "Flingern" diagram was established by the measurement of material losses at different heat exchanger tubes at an existing WtE plant. It is thus based on the corrosion mechanism occurring at the conditions of a WtE plant. In contrast, the calciner flue gas and the calciner fly ash show of a significantly different composition, which alters the occurring chemical reactions related to corrosion. Due to this fact and due to the highly complex and multilayer corrosion phenomena, this analysis is of rather qualitative nature and represents a first approach for the characterization of the risk of corrosion in the heat recovery system at a SRF oxyfuel-fired CaL calciner.

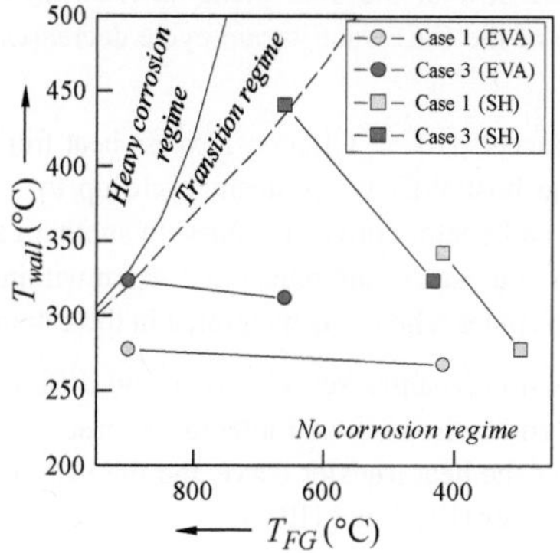

Figure 5-8: Corrosion regimes at heat exchanger surfaces in the calciner convective pass according to the "Flingern" diagram (back-end integration).

5.3.1.4 Advanced Heat Integration

Within this chapter, novel heat integration concepts (HIC) of high-temperature CaL process excess heat into the existing WtE water-steam cycle are proposed and discussed. The superheating and reheating of live steam is typically applied in large-scale power plants for improvements of the thermodynamic cycle efficiency, as it increases the mean temperature of heat addition [259]. It was already shown that external superheating and reheating is a promising approach for efficiency improvements of existing WtE plants [260, 261]. In the course of this assessment, two different HIC according to Figure 5-9 are considered.

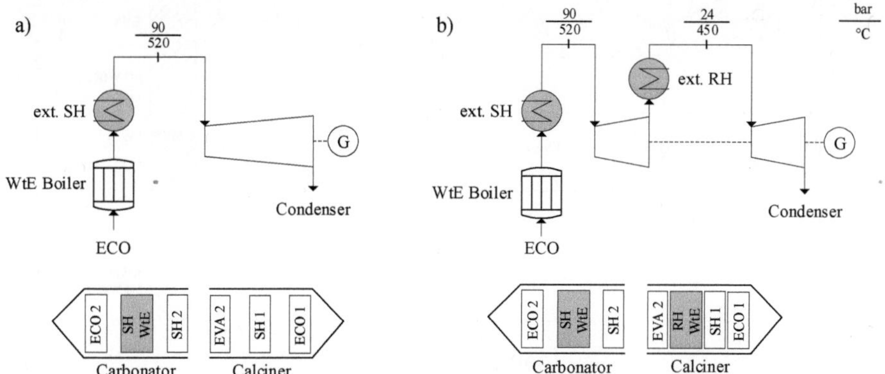

Figure 5-9: Advanced heat integration concepts (top), convective heat exchanger surfaces arrangement (bottom). HIC 2: External superheating (a), HIC 3: external superheating + reheating (b).

The sensible heat of the CO_2-depleted flue gas at the outlet of the carbonator and of the CO_2-product stream at the outlet of the calciner are utilized as additional heat source for the water-steam cycle of the WtE plant, respectively. Along with the temperature of the live steam, its pressure needs to be raised. Practically, this implies modifications of the existing WtE plant water-steam cycle. An economical evaluation is not performed here, because of large uncertainties regarding the costs for the required modifications. In each case, the thermal duty of the WtE boiler is kept constant. This implies that the feedwater mass flow of the WtE plant water-steam cycle is altered accordingly. Simultaneously, the dimension of the CaL water-steam cycle decreases as less heat is available for steam raising and superheating.

In the external superheating concept (HIC 2, Graph a), excess heat from the CaL process is used to superheat the live steam in the host WtE water-steam cycle up to 520 °C. Here, the WtE plant superheater is integrated in the carbonator convective pass. In another case (HIC 3, Graph b), excess heat from the CaL process is used to super- and reheat live steam within the WtE water-steam cycle (520 °C, 450 °C). Here, the WtE plant reheater is integrated in the calciner convective pass.

Table 5-15 summarizes the thermodynamic key results of the different HIC. As a reference, the corresponding numbers are given for the back-end integration base case (HIC 1) as well. Related to these results, Figure 5-10 shows the heat transfer curve and the heat distribution of the CaL water-steam cycle heat exchanger units for HIC 2 and HIC 3.

Table 5-15: Thermodynamic characteristics of heat integration scenarios.

Parameter	Unit	HIC 1 (Base)	HIC 2 (SH)	HIC 3 (RH + SH)
Total thermal input, $P_{th,tot}$	MW$_{th}$	132	132	132
Gross power output, $P_{el,gross,tot}$	MW$_e$	35.1	40.6	41.3
Net power output, $P_{el,net,tot}$	MW$_e$	21.0	26.6	27.3
CaL process gross power ratio, PR_{CaL}	%	55.2	43.4	39.9
Gross electrical efficiency, $\eta_{gross,tot}$	%	26.6	30.95	31.5
Net electrical efficiency, $\eta_{net,tot}$	%	16.0	20.3	20.8

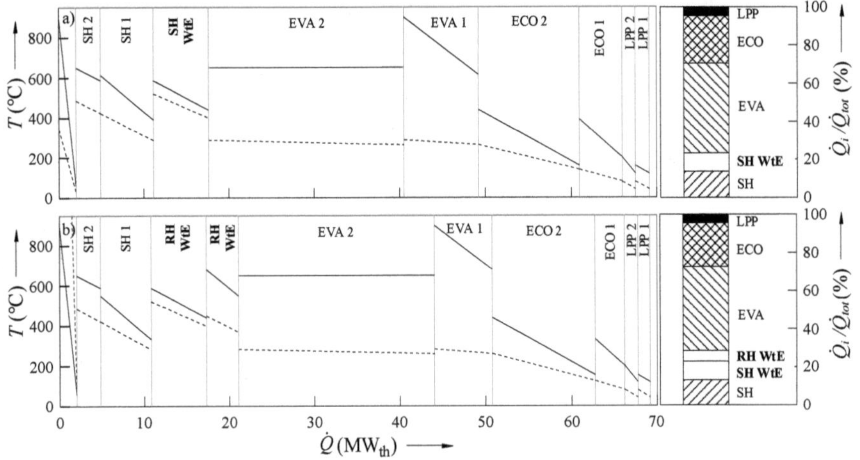

Figure 5-10: Heat transfer curve of the advanced heat integration options (left) and heat distribution (right). HIC 2: Superheating (a) and HIC 3: Reheating- and superheating (b).

With a simultaneous increase of the heat integration level, the CaL gross electrical ratio decreases while the overall net electrical efficiency increases. The gain in net electrical efficiency accumulates to more than 5 %-points for HIC 3. The major efficiency improvements are achieved by the superheating of WtE plant live steam. The additional gain in the net electrical efficiency by the reheating step is limited to approximately 0.5 %-points.

In case of HIC 2, approximately 9.7 % (6.5 MW$_{th}$) of the total CaL process excess heat is transferred to the WtE plant water steam cycle. In case of HIC 3, this numbers further increases to almost 15.3 % (10.3 MW$_{th}$), of which 5.7 % (3.8 MW$_{th}$) are related to the reheating step. As a consequence of the heat integration, the production of live steam in the CaL water-steam cycle decreases. In the base case, approximtely 20.4 kg/s of live steam are delivered to the inlet of the turbine, for HIC 2 (HIC 3) this number decreases to 18.4 kg/s (17.3 kg/s).

Figure 5-11 shows the corrosion regimes of heat exchanger surfaces in the calciner convective pass according to the "Flingern" diagram. Similarly to the previous assessment (see Figure 5-8), the process conditions of HIC 2 are fully located in the no corrosion regime. This is mainly due to the fact that the live steam parameters of the CaL water-steam cycle are moderate (485 °C, 60 bar). Furthermore, the required heat for external superheating is supplied by the CO_2 depleted flue gas at the outlet of the carbonator, which influences the conditions in the calciner heat recovery path only indirectly. During HIC 3, process conditions favor the risk of corrosion at the outlet of the external WtE plant reheater, as it is located at the transition regime. The limitation related to this corrosion characterization were already discussed in the previous analysis.

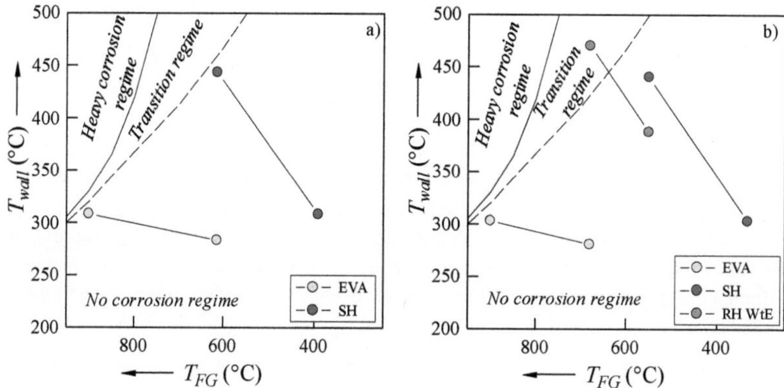

Figure 5-11: Corrosion regimes at heat exchanger surfaces in the calciner convective pass according to the "Flingern" diagram, HIC 2 (a) and HIC 3 (b).

5.3.1.5 Comprehensive Process Assessment

The comprehensive process assessment covers the complete set of parameters that were already addressed separately in the previous investigations. Herein, the superposition of all boundary conditions is discussed in terms of total net electrical efficiency ($\eta_{net,tot}$) and the SPECCA.

Figure 5-12 shows the total net electrical efficiency dependent on the CaL process gross power ratio for the different CaL process conditions, CaL water-steam cycle parameters and HIC. Additionally, the net electrical efficiency of the WtE plant without CaL process retrofit is given.

With regard to the total net electrical efficiency, the CaL process conditions and the CaL water-steam cycle parameters have a similar influence. Improved CaL process conditions and higher CaL water-steam cycle parameters increase the quantity of electricity that is deliverable per quantity of heat that is supplied to the retrofitted WtE plant. Controversially, in terms of CaL gross power ratio, the influence of both parameters differs. Once the CaL water-steam cycle parameters are raised, the available CaL process excess heat is converted more efficiently into electricity. Thus the CaL gross power ratio increases. In case of improved CaL process conditions, less heat needs to be supplied to the calciner by oxyfuel combustion. Consequently, less excess heat is available for steam generation and superheating, which lowers the CaL gross power ratio. The improvements of the net electrical efficiency in the latter case are mainly related to the reduced heat losses related to the first calcination of the make-up limestone and to the lower auxiliary power demand of the ASU and the GPU.

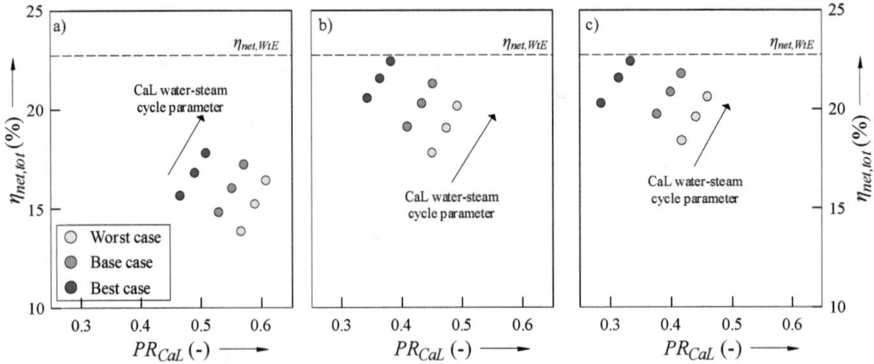

Figure 5-12: Comparison of total net electrical efficiency ($\eta_{net,tot}$) dependent on the CaL process gross power ratio (PR_{CaL}) for HIC 1 (a), HIC 2 (b) and HIC 3 (c).

While the total net electrical efficiency in the HIC 1 ranges below 18 % even under ideal CaL process conditions and CaL water-steam cycle parameters, this number increases close to the value of the WtE plant without CaL process retrofit for HIC 3. Here, the process performance at best CaL process conditions and highest CaL water-steam cycle parameters (520 °C / 80 bar), is thermodynamically limited by the available high-temperature heat for external superheating. Thus the production of live steam is limited and the gas streams of carbonator and calciner cannot be cooled down to the desired value of 120 °C.

Figure 5-13 shows the SPECCA dependent on the total net electrical power for the different CaL process conditions, CaL water-steam cycle parameters and HIC. The SPECCA are a function of the net electrical efficiency penalty and the specific CO_2 avoidance. It allows the direct comparison of CaL applications in the field of power plants. For coal fired power plants, this number ranges from 5.0 - 8.0 $MJ_{th}/kg_{CO2,av}$ [183, 202].

Overall, the SPECCA vary from 6.00 $MJ_{th}/kg_{CO2,av}$ (HIC 1, Worst case process conditions, 450 °C / 40 bar) to 0.70 $MJ_{th}/kg_{CO2,av}$ (HIC 3, Best case process conditions, 520 °C / 80 bar). For the baseline case in HIC 1, the efficiency drop is 7.53 %-points, which results in SPECCA of 4.56 $MJ_{th}/kg_{CO2,av}$. Despite this relatively high net electrical efficiency penalty, the SPECCA show a reasonable value, which is mainly due to biogenic CO_2 which allows for negative CO_2 emissions. In HIC 3, the SPECCA are below 2 $MJ_{th}/kg_{CO2,av}$ for best CaL process conditions and even below 1 $MJ_{th}/kg_{CO2,av}$ in case of the highest water-steam cycle parameters. Assuming these circumstances, one could state that the additional thermal efforts related to the avoidance of CO_2 are nearly insignificant.

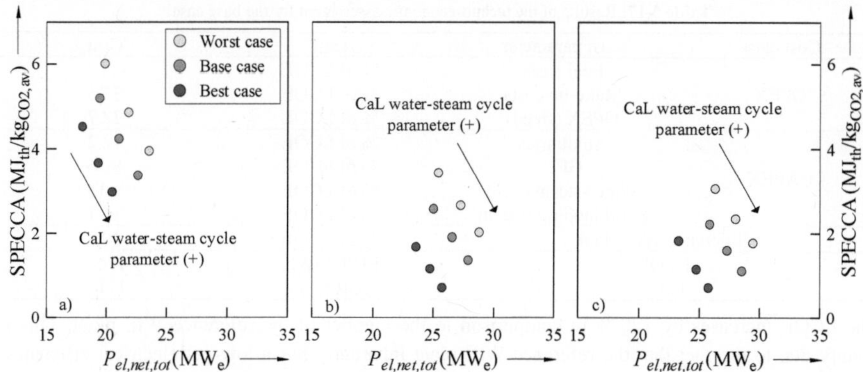

Figure 5-13: Comparison of SPECCA dependent on the total net power ($P_{el,net,tot}$) for HIC 1 (a), HIC 2 (b) and HIC 3 (c).

5.3.2 Economic Assessment

In the course of this chapter, the base case scenario and different techno-economic extreme scenarios are addressed with regard to relevant economic parameters, such as TPC, LCOE and CAC. Thereafter, the economics of the CaL process in the framework of WtE plants is discussed with respect to atmospheric CO_2 removal.

5.3.2.1 Economic Base Case Assessment

Table 5-16 shows the breakdown of the total TPC of CaL process components. The solid looping system comprises the two CFB reactors including the cyclone separators and all coupling components (e.g. loop seals, solid material pipes). The auxiliary units include the feeding systems for SRF and make-up limestone as well as the on-site ASU.

Table 5-16: Total plant costs (TPC) of CaL process components.

Component	Total plant costs MIO EUR	Share %
Solid looping system	47.6	35.3
Water-steam cycle	27.0	20.0
Auxiliaries	35.9	26.7
GPU	24.4	18.1
Total	135	100

Among the costs of the newly installed CO_2 capture unit, the expenses related to the solid looping system comprise the greatest share (35.3 %). This is partly related to the high material intensity of carbonator and calciner and to the high process contingency level that is assumed for this components. The costs associated with the water-steam cycle are moderate as a consequence of the relatively low live steam parameters which allow for the utilization of affordable piping steel.

Based on the annualized CAPEX and OPEX, the LCOE and CAC are calculated. Table 5-17 summarizes the corresponding numbers and the breakdown of LCOE for the WtE plant retrofitted by the CaL process. The TPC amount to 134.9 Mio EUR, whereas the OPEX accumulate to 6.9 Mio EUR/a.

Table 5-17: Results of the techno-economic assessment for the base case.

Cost class	Cost parameter	Unit	Cost
OPEX	Fuel costs	% of LCOE	-11.1
	Make-up costs	% of LCOE	3.35
	OPEX (fixed)	% of LCOE	32.7
CAPEX	Auxiliaries	% of LCOE	12.2
	GPU	% of LCOE	8.26
	Water-steam cycle	% of LCOE	9.15
	Solid looping system	% of LCOE	16.1
Reference WtE plant		% of LCOE	29.3
LCOE		EUR/MWh$_e$	176.2
CAC		EUR/t$_{CO2,av}$	111.8

The LCOE increases by 120 % in comparison to the number of the reference WtE plant. This is mainly due to the fact that the reference WtE plant inherently has a low net electrical efficiency, especially compared to large-scale power generation systems. Therefore, a relatively high amount of CO_2 needs to be captured per unit of electricity that is produced. Furthermore, the size of the WtE plant and of the CaL process is relatively small ($P_{th,WtE} \sim 60$ MW) compared to conventional power units ($P_{th} > 500$ MW), which are typically considered for CCS applications. Consequently, the benefits due to the economics of size are limited. As a result of the high LCOE increase, CAC above 111.8 EUR/t$_{CO2,av}$ are obtained.

The OPEX of the CaL system account for 23.2 % of the total LCOE, whereas the CAPEX of the CaL system accumulate to 49.6 % of the total LCOE. The negative price of SRF significantly improves the system economics as the LCOE are reduced by 19.5 EUR/MWh$_e$.

While the costs presented above may seem high, it is important to realize that by considering CCS in the framework of a WtE plant, it enables a new business opportunity in the removal of atmospheric CO_2 due to the organic waste fractions in MSW and SRF. The economics of the retrofitted WtE plant are further discussed in Chapter 5.3.2.3, while taking this aspect into account.

5.3.2.2 Analysis of Economical Extreme Scenarios

In the course of this chapter, economic extreme scenarios are evaluated to account for uncertainties related to the boundary conditions in the course of the economic evaluation. Table 5-18 summarizes the boundary conditions for the worst, base and best case scenario.

Table 5-18: Boundary conditions for the economic evaluation of extreme scenarios.

Parameter	Unit	Worst Case	Base Case	Best case
Utilization factor	%	75	85	95
Interest rate	%	10	8	6
SRF price	EUR/t	0	-25	-50
Make-up price	EUR/t	15	10	5
Plant lifetime	A	20	25	30

As already noted, the price of SRF is highly case specific and varies among countries and regions. A relatively low or even negative fuel price is essential to make the CaL process retrofit to WtE plants an economically competitive option. Among the economic extreme scenarios the SRF price is varied between 0 EUR/t and -50 EUR/t. Furthermore, the interest rate and the plant lifetime are modified, which alters the costs related to the annualized investment costs.

Figure 5-14 shows the breakdown of LCOE and the resulting CAC of the extreme cases in addition to the numbers referring to the base case.

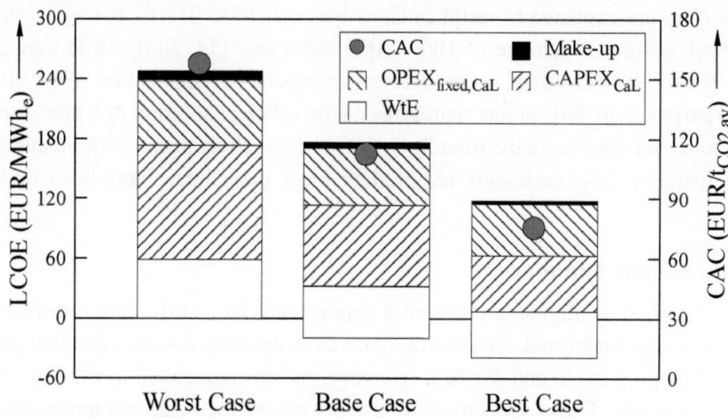

Figure 5-14: LCOE and CAC for the economic extreme case scenarios.

Under economically worst conditions, the LCOE increase to 247.2 EUR/MWh$_e$, which leads to CAC of 156.9 EUR/t$_{CO2,av}$. Contrary, the optimistic assumption of the best case reduces LCOE and CAC to 117.6 EUR/MWh$_e$ and 74.6 EUR/t$_{CO2,av}$. The major cause for this improvement are the revenues for SRF treatment, which accumulate to 6.85 MIO EUR/a in the best case scenario. It needs to be noted that even under unfavorable assumptions, the CaL process represents an economically favorable option for CO_2 capture from WtE plants in comparison to MEA-based CO_2 capture (LCOE: 342 EUR/MWh$_e$, CAC: 217 EUR/t$_{CO2,av}$) and advanced solvent-based CO_2 capture (LCOE: 279 EUR/MWh$_e$, CAC: 153 EUR/t$_{CO2,av}$) [252].

5.3.2.3 Economics with Respect to Atmospheric CO_2 Removal

In addition to the thermal treatment of wastes and the delivery of power and/or heat, the supply of negative CO_2 emissions represents another product of a CaL equipped WtE plant in the future. Figure 5-15 shows the LCOE as a function of the credit for CO_2 avoidance.

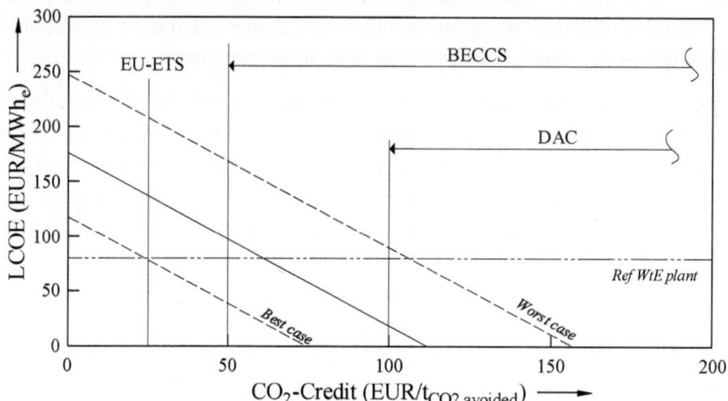

Figure 5-15: LCOE dependent on credit for avoided CO_2.

Economics are valorized once credits for negative CO_2 emissions are sold to other market actors to compensate for their CO_2 emissions. Even though, the cost range is still high compared to the current EU ETS price (25 EUR/t$_{CO2,av}$) [262], it is on the low end when compared to some of the other means considered to enable negative CO_2 emissions, such as typical BECCS or DAC systems. As shown in

Figure 5-15, BECCS are expected to result in CAC between 50 to 250 EUR/$t_{CO2,av}$ [263] and DAC are hoped to reach costs in the range of 100 - 200 EUR/$t_{CO2,av}$ [54, 264], while they are currently being sold for 980 EUR/$t_{CO2,av}$ [265]. Considering this aspect, the application of an SRF-fired CaL process in the framework of WtE plants represents a cost-efficient solution to while at the same time treating MSW in an environmentally friendly way, produce clean power and enable a significant amount of the negative CO_2 emissions required to meet the climate ambitions under the Paris Agreement.

5.4 Techno-Economic Summary

In this Chapter, a techno-economic analysis of a generic 60 MW_{th} WtE plant retrofitted by a SRF-fueled CaL process was performed. The thermodynamic modelling covers a detailed process model for the CaL solid looping cycle and the heat recovery systems associated to the WtE plant and the CaL process, respectively. The process modelling work covers the back-end integration of the CaL process, as the standard approach. In addition to that, novel heat integration concepts are proposed and thermodynamically assed. This includes the external super- and reheating of WtE live steam by high-temperature CaL process excess heat. Investment costs were calculated using a bottom-up approach considering the costs for each component. The techno-economic assessment is carried out based on KPI relevant for CCS process applications in the field of power production facilities. Additionally, the influence of relevant boundary conditions made in the course of the techno-economic assessment on the KPI is addressed by means of a dedicated sensitivity analysis.

In the base case retrofit scenario, the net electrical output of the retrofitted WtE plant increases from 13.6 MW_e up to 21.0 MW_e while the electrical efficiency decreases from 22.7 % to 16.0 %. Here, the specific CO_2 emissions accumulate to -925 g_{CO2}/kWhe, which indicates a strongly negative CO_2 emission balance. In a sensitivity analysis, the influence of CaL process conditions and different excess heat integration options were assessed. In case of a proper CaL excess heat integration into the existing WtE power cycle by means of external super- and reheating, the net electrical efficiency could be raised by more than 5 %-points compared to the reference WtE plant without CaL process. Based on the "Flingern" diagram, it was proven that the calciner convective pass could be operated outside the heavy corrosion regime throughout all heat integration options.

In the course of the economic process assessment, it was found that the LCOE increases significantly from 80 EUR/MWh_e (WtE plant w/o CaL process) to 176 EUR/MWh_e (WtE plant w/ CaL process). In contrast to large-scale CaL integration studies, the small thermal size of the WtE plant and low electrical efficiency of the reference WtE plant are the main cause for this increase, as a relatively large quantity of CO_2 needs to be captured per quantity of electricity that is supplied. Based on the increase in LCOE and the specific costs for CO_2 avoidance yield to 112 EUR/$t_{CO2,av}$. While the derived costs may seem high, it is important to realize that by considering CCS in the framework of a WtE plant, it enables a new business opportunity in the removal of atmospheric CO_2 due to the organic waste fractions in MSW and SRF. In this regard, CaL process equipped WtE plants clearly represents a competitive solution in comparison with other technologies that allow for negative CO_2 emissions.

6 Conclusions and Further Work

Within the course of this thesis, the utilization of waste-derived fuels in the CaL process was successfully demonstrated at 1 MW_{th} scale for the first time worldwide. Furthermore, a CaL process model was implemented, validated and applied for the calculation of heat and mass flows for a CaL process retrofit onto a WtE plant. Based on a comprehensive thermodynamic assessment, a techno-economic evaluation reveals the costs related to CO_2 avoidance in the framework of WtE plants.

By more than 230 hours of representative CaL operation at the 1 MW_{th} CaL pilot plant, the continuous CO_2 capture was successfully demonstrated. Carbonator CO_2 absorption rates close to the chemical equilibrium and total CO_2 capture rates over 90 % were experimentally proven. Throughout all test points, the carbonator was operated without active reactor cooling. This was feasible as less CO_2 was present in the flue gas to be decarbonized in comparison to the exhaust gases of coal-fired power plants or cement plants. The evaluation of process performance bases on the closure of carbon balances for the carbonator and for the calciner. Exemplarily, it was found that a carbonator active space time of 60 s is required to achieve a nominalized CO_2 absorption efficiency above 80 %. The analysis of the calciner reveals that a calcination temperature of 840 °C is sufficient for a calcination efficiency above 90 %, as the CO_2 partial pressure in the riser is reduced by the presence of water vapor, nitrogen and oxygen.

In the course of a comprehensive solid sample analysis, several agreements between process conditions at the 1 MW_{th} pilot plant and the properties of the sorbent samples were found. Exemplarily, the activity of the sorbent increases along with a low calcination temperature, while sorbent deactivation is facilitated in case of a low limestone make-up feed. Furthermore, the reduction of the mean particle size is in agreement with the development of the sorbent attrition coefficient. These findings state the readiness of the technology for further scale-up as it proves the accuracy of the applied process evaluation tools and methodologies.

Part of the calciner evaluation is associated with the FB combustion of SRF with regard to major gaseous emission pollutants such as CO, NO, SO_2 and HCl. Here, the combustion in air, oxygen-enriched air and in a real oxyfuel environment was evaluated for the CaL calciner and for the combustion in a stand-alone CFB with a silicate sand inventory. For both types of SRF, the combustion in four distinguished regimes was thus assessed. The overall level of specific CO emissions was relatively high throughout all experimental investigations, which indicates a poor fuel burnout. This is associated with a fluctuating fuel mass flow and a poor mixing of volatile species and oxidant. Both phenomena can be significantly improved by an appropriate design at a large-scale unit. The formation tendency of NO reveals that a higher oxygen-to-fuel ratio and higher nitrogen fraction in the fuel support the formation of NO. The oxyfuel combustion environment lowers the emissions of NO. The specific emissions of SO_2 during SRF combustion in a silicate sand inventory varied between 20 and 70 mg_{SO2}/MJ_{th}, while the ideal temperature for sulfation was approximately 850 °C. However, the specific emissions of SO_2 during the SRF combustion in the CaL calciner were very low and close to the detection limit of the gas analyzer throughout the experimental investigations. The emission characteristics of HCl were also greatly influenced by the properties of the solid phase. Accordingly, Cl retention rates above 80 % were achieved during the course of oxyfuel combustion in the CaL calciner. Contrary, the Cl retention decreases to 18 - 58 % during oxyfuel combustion in a silicate sand CFB.

A CaL process model was applied for the prediction of CaL process performance. As the first step, a model validation was carried out in terms of a comparison of the CO_2 absorption efficiency derived from experimental data and from the process model. It was found that the majority of validation points (75 %) were calculated with an uncertainty error of +/- 10 %. The validated process model was subsequently applied to the calculation of heat and mass flows in a retrofit scenario on a 60 MW_{th} WtE plant. Based on a thermodynamic process evaluation, a techno-economic assessment was carried out. Additionally, the influence of relevant boundary conditions made in the course of the techno-economic assessment on the KPI's was addressed by means of a dedicated sensitivity analysis. It was found that the LCOE increases significantly from 80 EUR/MWh_e (WtE plant w/o CaL process) to 176 EUR/MWh_e (WtE plant w/ CaL process). Based on the increase in LCOE and the specific CO_2 avoidance, the CAC yields up to 112 EUR/$t_{CO2,av}$. In contrast to large-scale CaL process integration studies, the small thermal size of the WtE plant and the low electrical efficiency of the reference WtE plant are the main cause for this increase, as a relatively large quantity of CO_2 needs to be captured per quantity of electricity that is supplied. Ultimately, it was found that WtE plants retrofitted by a SRF-fueled CaL process for CO_2 capture can compete with BECCS systems or other technologies to provide cost-efficient negative CO_2 emissions.

Further research should address the effect of HCl on the sorbent activity. Here, also the presence of HCl in the flue gas to be decarbonized in the carbonator needs to be considered. There is the potential to capture CO_2 and HCl simultaneously in the carbonator, thus relieving parts of the flue gas cleaning line of the host WtE plant. Another aspect is related to the combustion of SRF in the calciner. Further experimental investigations are required to fully assess the combustion behavior and the gaseous emission mitigation potential by means of primary measures. This includes variations in terms of combustion temperature, secondary air staging, oxygen concentration in the oxidation agent and oxygen-to-fuel ratio. Hereby, it is key to balance the efforts for CO_2-product purification and the efforts related to the actual combustion process, while continuously ensuring a sufficient sorbent regeneration in the calciner. The most important step in the further development of this technology is the design and the operation of a demonstration plant in the thermal scale of approximately 20 MW_{th}. Doing so, a comprehensive data basis could be established that allows a more accurate predication of CO_2 capture behavior and the related techno-economic process performance.

Bibliography

[1]   S. Solomon, G.-K. Plattner, R. Knutti and P. Friedlingstein. Irreversible climate change due to carbon dioxide emissions. *Proceedings of the National Academy of Sciences of the United States of America*, 106(06): 1704-1709, 2009. DOI: 10.1073/pnas.0812721106.

[2]   S. Manabe and R.J. Stouffer. Sensitivity of a global climate model to an increase of CO_2 concentration in the atmosphere. *Journal of Geophysical Research*, 85(C10):5529-5554, 1980. DOI: 10.1029/JC085iC10p05529.

[3]   H. Rohde. A comparison of the contribution of various gases to the greenhouse effect. *Science*, 248(4960): 1217-1219, 1990. DOI: 10.1126/science.248.4960.1217.

[4]   International Energy Agency (IEA). *World Energy Outlook 2019: Executive Summary*. 2019.

[5]   D.M. Etheridge, L.P. Steele, R.L. Langenfelds, R.J. Francey, J.-M. Barnola and V.I. Morgan. Natural and anthropogenic changes in atmospheric CO_2 over the last 1000 years from air in Antarctic ice and firn. *Journal of Geophysical Research*, 101(D2): 4115-4128, 1996. DOI: 10.1029/95JD03410.

[6]   C.D. Keeling, S.C. Piper, R.B. Bacastow, M. Wahlen, T.P. Whorf, M. Heimann and H.A. Meijer. Atmospheric CO_2 and $13CO_2$ exchange with the terrestrial biosphere and oceans from 1978 to 2000: Observation and carbon cycle implications. *Ecological Studies (Analysis and Synthesis)*, 177. Springer, New York, NY.

[7]   International Energy Agency (IEA). CO_2 Emission statistics. 2018. URL: https://www.iea.org/subscribe-to-data-services/co2-emissions-statistics. Cited: 18.02.2020.

[8]   United Nations Climate Change. *Historic Paris Agreement on Climate Change: 195 Nations Set Path to Keep Temperature Rise Well Below 2 Degrees Celsius*. URL: https://unfccc.int/news/finale-cop21. Cited: 15.02.2019.

[9]   International Energy Agency (IEA). *Energy Technology Perspectives*. 2017.

[10]   D.A. Hoornweg and P. Bhada-Tata. *What a waste: A global review of solid waste management*. World Bank Group, Washington, 2012.

[11]   H.-J. Gehrmann, M. Hiebel and F.-G. Simon. Methods for the evaluation of waste treatment processes. *Journal of Engineering*, 3567865:13, 2017. DOI: 10.1155/2017/3567865.

[12]   D.C. Wilson, L. Rodic, P. Modak, R. Soos, A.C. Rogero, C. Velis and M. Iyer. *Global Waste Management Outlook*. 2015.

[13]   European Commission. *The role of waste-to-energy in the circular economy*. 2017.

[14]   United States Environmental Protection Agency (EPA). *Solid waste management and greenhouse gases - a life-cycle assessment of emissions and sinks*. 2006.

[15]   S. Flamme and J. Geiping. Quality standards and requirements for solid recovered fuels: a review. *Waste Management & Research*, 30(4): 335-353, 2012. DOI: 10.1177/0734242X12440481.

[16]   M. Schneider. Process technology for efficient and sustainable cement production. *Cement and Concrete Research*, 78(A): 14-23, 2015. DOI: 10.1016/j.cemconres.2015.05.014.

[17]   T. Hilber, J. Maier, G. Scheffknecht, M. Agraniotis, P. Grammelis, E. Kakaras, T. Glorius, U. Becker, W. Derichs, H.-P. Schiffer, M.D. Jong and L. Torri. Advantages and possibilities of solid recovered fuel combustion in the European energy sector. *Journal of Air & Waste Management Association*, 57(10): 1178-1189, 2007. DOI:10.3155/1047-3289.57.10.1178.

[18]   D. Lee, J.s. Shin, J.h. Park, D.-H. Bae and D. Shun. Performance evaluation of a 60 ton/h CFB boiler for solid refuse fuel. *Proceedings of the 23rd International Conference on FBC*, Seoul (South Korea), 2018.

[19]   M.C. Di Lonardo, M. Franzese, G. Costa, R. Gavasci and F. Lombardi. The application of SRF vs. RDF classification and specifications to the material flows of two mechanical-biological treatment plants of Rome: Comparison and implications. *Waste Management*, 47(B): 195-205, 2016. DOI: 10.1016/j.wasman.2015.07.018.

[20]   CEN, EN 15358, Solid recovered fuels - Specification and classes. Brussels: CEN.

[21]   A. Garg, R. Smith, D. Hill, P.J. Longhurst, S.J.T. Pollard and N. Simms. An integrated appraisal of energy recovery options in the United Kingdom using solid recovered fuels derived from municipal solid waste. *Waste Management*, 29(8): 2289-2297, 2009. DOI: 10.1016/j.wasman.2009.03.031.

[22]   R. Anantharaman, O. Bolland, N. Booth, E. van Dorst, C. Ekstrom, E. Sanchez Fernandes, F. Franco, E. Macchi, G. Manzolini, D. Shell, A. Pfeffer, M. Prins, S. Rezvani and L. Robinson. *CAESAR Deliverable 4.9: European best practice guideline for assessment of CO_2 capture technologies.* 2011.

[23]   S.V. Vassilev, D. Baxter, L.K. Andersen and C.G. Vassileva. An overview of the chemical composition of biomass. *Fuel*, 89(5): 913-933, 2010. DOI:10.1016/j.fuel.2009.10.022.

[24]   M. Haaf, R. Anantharaman, S. Roussanaly, J. Ströhle and B. Epple. CO_2 capture from waste-to-energy plants: Techno-economic assessment of novel integration concepts of calcium looping technology. *Resources, Conservation & Recycling*, 2020. (accepted for publication).

[25]   P.H. Brunner and H. Rechberger. Waste to energy - key element for sustainable waste management. *Waste Management*, 37: 3-12, 2015. DOI: 10.1016/j.wasman.2014.02.003.

[26]   Eurostat. *Municipal waste landfilled, incinerated, recycled and composted.* URL: https://ec.europa.eu/eurostat/statistics-explained/index.php?title=File:Municipal_waste_landfilled,_incinerated,_recycled_and_composted,_EU-28,_1995-2017.png. Cited: 15.02.2020.

[27]   P. Viklund, A. Hjörnhede, P. Henderson, A. Stalenheim and R. Pettersson, Corrosion of superheater materials in a waste-to-energy plant. *Fuel Processing Technology*, 105: 106-112, 2013. DOI: 10.1016/j.fuproc.2011.06.017.

[28]   H.P. Nielsen, F.J. Frandsen, K. Dam-Johansen and L.L. Baxter. The implications of chlorine-associated corrosion on the operation of biomass-fired boilers. *Progress in Energy and Combustion Science*, 26(3): 283-298, 2000. DOI: 10.1016/S0360-1285(00)00003-4.

[29]   E. Vainio, H. Kinnunen, T. Lauren, A. Brink, P. Yrjas, N. DeMartini and M. Hupa. Low-temperature corrosion in co-combustion of biomass and solid recovered fuels. *Fuel*, 184: 957-965, 2016. DOI:10.1016/j.fuel.2016.03.096.

[30]   Österreichischer Wasser- und Abfallwirtschaftsverband (ÖWAV). ÖWAV-Regelblatt 519: *Energetische Wirkungsgrade von Abfallverbrennungsanlagen.* Wien, 2013.

[31]   J. Dong, Y. Tang, A. Nzihou, Y. Chi, E. Weiss-Hortala, M. Ni and Z. Zhou. Comparison of waste-to-energy technologies of gasification and incineration using life cycle assessment: Case studies in Finland, France and China. *Journal of Cleaner Production*, 203: 287-300, 2018. DOI: 10.1016/j.jclepro.2018.08.139.

[32]   L. Lombardi, E. Carnevale and A. Corti. A review of technologies and performance of thermal treatment systems for energy recovery from waste. *Waste Management*, 37: 26-44, 2015. DOI: 10.1016/j.wasman.2014.11.010.

[33]   European Commission. *Best Available Techniques (BAT) Reference Document for Waste Incineration - Final Draft.* 2018.

[34]   J. Gibbins and H. Chalmers. Carbon Capture and storage. *Energy Policy*, 36(12): 4317-4322, 2008. DOI: 10.1016/j.enpol.2008.09.058.

[35]   Intergovernmental Panel on Climate Change (IPCC). *IPPC Special Report on Carbon Dioxide Capture and Storage.* 2005.

[36]   E. Billig, M. Decker, W. Benzinger, F. Ketelsen, P. Pfeifer, R. Peters, D. Stolten and D. Thrän. Non fossil CO_2 recycling - the technical potential for the present and future utilization for

fuels in Germany. *Journal of CO_2 Utilization*, 30: 130-141, 2019. DOI: 10.1016/j.jcou.2019.01.012.

[37] J.C. Abanades, E.S. Rubin, M. Mazzotti, and H.J. Herzog. On the climate change mitigation potential of CO_2 conversion to fuels. *Energy & Environment Science*, 10(12): 2491-2499, 2017. DOI: 10.1039/c7ee02819a.

[38] W.-H. Chen, S.-M. Chen and C.-I. Hung. Carbon dioxide capture by single droplet using Selexol, Rectisol and water as absorbents: A theoretical approach. *Applied Energy*, 111: 731-741, 2013. DOI: 10.1016/j.apenergy.2013.05.051.

[39] C. Descamps, C. Bouallou and M. Kanniche, Efficiency of an integrated gasification combined cycle (IGCC) power plant including CO_2 removal. *Energy*, 33(6): 874-881, 2008. DOI:10.1016/j.energy.2007.07.013.

[40] C. Kunze and H. Spliethoff. Modelling of an IGCC plant with carbon capture for 2020. *Fuel Processing Technology*, 91(8): 934-941, 2010. DOI:10.1016/j.fuproc.2010.02.017.

[41] D.Y.C. Leung, G. Caramanna, and M.M. Maroto-Valer, An overview of status of carbon dioxide capture and storage technologies. *Renewable and Sustainable Energy Reviews*, 39: 426-443, 2014. DOI: 10.1016/j.rser.2014.07.093.

[42] H.C. Mantripragada, H. Zhai and E.S. Rubin. Boundary Dam or Petra Nova - which is a better model for CCS energy supply. *International Journal of Greenhouse Gas Control*, 82: 59-68, 2019. DOI: 10.1016/j.ijggc.2019.01.004.

[43] K. Stèphenne. Start-up of world's first commercial post-combustion coal fired CCS project: Contribution of shell cansolv to SaskPower Boundary Dam ICCS project. *Energy Procedia*, 63: 6106-6110, 2014. DOI: 10.1016/j.egypro.2014.11.642.

[44] D.P. Hanak, C. Biliyok and V. Manovic. Efficiency improvements for the coal-fired power plant retrofit with CO_2 capture plant chilled ammonia process. *Applied Energy*, 151(1):258-272, 2015. DOI: 10.1016/j.apenergy.2015.04.059.

[45] V. Darde, K. Thomsen, W.J.M. van Well and E.H. Stenby. Chilled ammonia process for CO_2 capture. *International Journal of Greenhouse Gas Control*, 4(2):131-136, 2010. DOI:10.1016/j.ijggc.2009.10.005.

[46] R. Bredesen, K. Jordal and O. Bolland. High-temperature membranes in power generation with CO_2 capture. *Chemical Engineering and Processing*, 43(9): 1129-1158, 2004. DOI:10.1016/j.cep.2003.11.011.

[47] M.-B. Hägg and A. Lindbrathen. CO_2 capture from natural gas fired power plants by using membrane technology. *Industrial Engineering & Chemistry Research*, 44(20): 7668-7675, 2005. DOI: 10.1021/ie050174v.

[48] N. Prosser and M. Shah. Air separation Units for Oxy-coal Power Plants. *Proceedings of the 2nd Oxyfuel Combustion Conference*, Queensland (Australia), 2011.

[49] R. Stanger, T. Wall, R. Spörl, M. Paneru, S. Grathwohl, M. Weidmann, G. Scheffknecht, D. McDonald, K. Myöhänen, J. Ritvanen, S. Rahiala, T. Hyppänend, J. Mletzko, A. Kather and S. Santos. Oxyfuel combustion for CO_2 capture in power plants. *International Journal of Greenhouse Gas Control*, 40: 55-125, 2015. DOI: 10.1016/j.ijggc.2015.06.010.

[50] T. Fujimori and T. Yamada. Realization of oxyfuel combustion for near zero emission power generation. Proceedings of the Combustion Institute, 34(2): 2111-2130, 2013. DOI: 10.1016/j.proci.2012.10.004

[51] J. Adánez, A. Abad, T. Mendiara, P. Gayán, L.F de Diego and F. Garcìa-Labiano. Chemical looping combustion of solid fuels. *Progress in Energy and Combustion Science*, 65: 6-66, 2018. DOI: 10.1016/j.pecs.2017.07.005.

[52] T. Mattisson, M. Keller, C. Linderholm, P. Moldenhauer, M. Rydèn, H. Leion and A. Lyngfelt. Chemical-looping technologies using circulating fluidized bed systems: Status of development. *Fuel Processing Technology*, 172: 1-12, 2018. DOI: 10.1016/j.fuproc.2017.11.016.

[53] J.C. Abanades, B. Arias, A. Lyngfelt, T. Mattisson, D.E. Wiley, H. Li, M.T. Ho, E. Mangano and S. Brandani. Emerging CO_2 capture systems. *International Journal of Greenhouse Gas Control*, 40:126-166, 2015. DOI: 10.1016/j.ijggc.2015.04.018.

[54] M. Bui, C.S. Adjiman, A. Bardow, E.J. Anthony, A. Boston, S. Brown, P.S. Fennell, S. Fuss, A. Galindo, L.A. Hackett, J.P. Hallett, H.J. Herzog, G. Jackson, J. Kemper, S. Krevor, G.C. Maitland, M. Matuszewski, I.S. Metcalfe, C. Petit, G. Puxty, J. Reimer, D.M. Reiner, E.S. Rubin, S.A. Scott, N. Shah, B. Smit, J.P.M. Trusler, P. Webley, J. Wilcox and N. Mac Dowell. Carbon capture and storage (CCS): the way forward. *Energy & Environmental Science*, 11: 1062-1176, 2018. DOI: 10.1039/C7EE02342A.

[55] International Energy Agency Greenhous Gas R&D Programme (IEAGHG). *Further assessment of emerging CO_2 capture technologies for the power sector and their potential to reduce costs.* 2019.

[56] Z. Abbas, T. Mezher and M.R.M. Abu-Zahra. CO_2 purification. Part I: Purification requirement review and the selection of impurities deep removal technologies. *International Journal of Greenhouse Gas Control*, 16: 324-334, 2013. DOI: 10.1016/j.ijggc.2013.01.053.

[57] H. Rütters, S. Stadler, R. Bäßler, D. Bettge, S. Jeschke, A. Kather, C. Lempp, U. Lubenau, C. Ostertag-Henning, S. Schmitz, S. Schütz, S. Waldmann and the COORAL Team. Towards an optimization of the CO_2 stream composition - A whole chain approach. *International Journal of Greenhouse Gas Control*, 54(2): 682-701, 2016. DOI:10.1016/j.ijggc.2016.08.019.

[58] S. Posch and M. Haider. Optimization of CO_2 compression and purification units (CO_2CPU) for CCS power plants. *Fuel*, 101: 254-263, 2011. DOI:10.1016/j.fuel.2011.07.039.

[59] E. de Visser, C. Hendriks, M. Barrio, M.J. Mølnvik, G. de Koeijer, S. Liljemark and Y. Le Gallo. Dynamis CO_2 quality recommendations. *International Journal of Greenhouse Gas Control*, 2(4): 478-484, 2008. DOI: 10.1016/j.ijggc.2008.04.006.

[60] S. Bachu. Review of CO_2 storage efficiency in deep saline aquifers. *International Journal of Greenhouse Gas Control*, 40: 188-202, 2015. DOI: 10.1016/j.ijggc.2015.01.007.

[61] R.A. Chadwick, P. Zweigel, U. Gregersen, G.A. Kirby, S. Holloway and P.N. Johannessen. Geological reservoir characterization of a CO_2 storage site: The Utsira Sand, Sleipner, northern North Sea. *Energy*, 29(9-10): 1371-1381, 2004. DOI: 10.1016/j.energy.2004.03.071.

[62] A. Singh and K. Stéphenne. Shell Cansolv CO_2 capture technology: Achievement from First Commercial Plant. Energy Procedia, 63: 1678-1685, 2014. DOI: 10.1016/j.egypro.2014.11.177.

[63] J.Q. Shi and S. Durucan. CO_2 storage in deep unminable coal seams. *Oil & Gas Science and Technology*, 60(3): 547-558, 2005. DOI: 10.2516/ogst:2005037.

[64] International Energy Agency (IEA). *Combining Bioenergy with CCS.* 2011.

[65] J. Kemper. Biomass and carbon dioxide capture and storage: A review. *International Journal of Greenhouse Gas Control*, 40: 401-430, 2015. DOI: 10.1016/j.ijggc.2015.06.012.

[66] M. Fajardy and N. Mac Dowell. Can BECCS deliver sustainable and resource efficient negative emissions? *Energy & Environment Science*, 10: 1389-1426, 2017. DOI: 10.1039/C7EE00465F

[67] C. Gough, S. Garcia-Freites, C. Jones, S. Mander, B. Moore, C. Pereira, M. Röder, N. Vaughan and A. Welfe. Challenges to the use of BECCS as a keystone technology in pursuit of 1.5 C. *Global Sustainability*, 1(e5): 1-9, 2018. 1. DOI: 10.1017/sus.2018.3.

[68] S. Pour, P.A. Webley and P.J. Cook, Potential for using municipal waste as a resource for bioenergy with carbon capture and storage (BECCS). *International Journal of Greenhouse Gas Control*, 68:1-15, 2018. DOI:10.1016/j.ijggc.2017.11.007.

[69] M. Haaf, P. Ohlemüller, J. Ströhle and B. Epple, Techno-economic assessment of alternative fuels in second-generation carbon capture and storage processes. *Mitigation and Adaptation Strategies for Global Change*, 25:149-164 2020. DOI: 10.1007/s11027-019-09850-z.

[70] P. Smith, S.J. Davis, F. Creutzig, S. Fuss, J. Minx, B. Gabrielle, E. Kato, R.B. Jackson, A. Cowie, E. Kriegler, D.P. van Vuuren, J. Rogelj, P. Ciais, J. Milne, J.G. Canadell, D.

McCollum, G. Peters, R. Andrew, V. Krey, G. Shrestha, P. Friedlingstein, T. Gasser, A. Grübler, W.K. Heidug, M. Jonas, C.D. Jones, F. Kraxner, E. Littleton, J. Lowe, J.R. Moreira, N. Nakicenovic, M. Obersteiner, A. Patwardhan, M. Rogner, E.S. Rubin, A. Sharifi, A. Torvanger, Y. Yamagata, J. Edmonds and C. Yongsung. Biophysical and economic limits to negative CO_2 emissions. *Nature Climate Change*, 6: 42-50, 2016. DOI: 10.1038/nclimate2870.

[71] F. Zeman. Energy and Material Balance of CO_2 Capture from Ambient Air. *Environment Science Technology*, 41(21): 7558-7563, 2007. DOI: 10.1021/es070874m.

[72] A. Goeppert, M. Czaun, G.K.S. Prakash and G.A. Olah. Air as the renewable carbon source of the future: an overview of CO_2 capture from the atmosphere. *Energy & Environmental Science*, 5: 7833-7853, 2012. DOI: 10.1039/C2EE21586A.

[73] J.C.M. Pires. Negative emissions technologies: A complementary solution for climate change mitigation. *Science of the Total Environment*, 672: 502-514, 2019. DOI:10.1016/j.scitotenv.2019.04.004.

[74] J. Strefler, T. Amann, N. Bauer, E. Kriegler, and J. Hartmann. Potential and costs of carbon dioxide removal by enhanced weathering of rocks. *Environmental Research Letters*, 13(3): 034010, 2018. DOI: 10.1088/1748-9326/aaa9c4.

[75] European Academics Science Advisory Council (easac). *Negative emission technologies: What role in meeting Paris Agreement targets?* 2018.

[76] R.B. Jackson, J.T. Randerson, J.G. Canadell, R.G. Anderson, R. Avissar, D.D. Baldocchi, G.B. Bonan, K. Caldeira, N.S. Diffenbaugh, C.B. Field, B.A. Hungate, E.G. Jobbàgy, L.M. Kueppers, M. De Nosetto and D.E. Pataki. Protecting climate with forests. *Environmental Research Letters*, 3: 044006, 2008. DOI: 10.1088/1748-9326/3/4/044006.

[77] D.J. Farrelly, C.D. Everard, C.C. Fagan and K.P. McDonnell. Carbon sequestration and the role of biological carbon mitigation: A review. *Renewable and Sustainable Energy Reviews*, 21: 712-727, 2013. DOI: 10.1016/j.rser.2012.12.038.

[78] D.T. Kearns. *Waste-to-Energy with CCS: A pathway to carbon-negative power generation.* Global CCS Institute, 2019.

[79] J. Stuen. Feasibility Study of Capturing CO_2 from the Kelmetsrud CHP Waste-to-Energy Plan in Oslo. *Energie aus Abfall*, 14: 133-144, 2017.

[80] The Chemical Engineer. *Aker Solutions to provide carbon capture technology to waste-to-energy plant.* 2019. URL: https://www.thechemicalengineer.com/news/aker-solutions-to-provide-carbon-capture-technology-to-waste-to-energy-plant/. Cited: 20.12.2019.

[81] Global CCS Institute. Saga City: The world's best kept secret (for now). 2018. URL: https://www.globalccsinstitute.com/news-media/insights/saga-city-the-worlds-best-kept-secret-for-now/. Cited: 20.12.2019.

[82] J. Kremer. *Analyse der Entwicklungsstufen des Carbonate Looping Prozesses.* Ph.D. Thesis, Technische Universität Darmstadt, 2014.

[83] D. Kunii and O. Levenspiel. *Fluidization Engineering.* Boston: Butterworth-Heineman, 2nd edition, 1991.

[84] D. Geldart. *Gas Fluidization Technology.* John Wiley and Sons, 1986.

[85] M. Kraume. *Transportvorgänge in der Verfahrenstechnik: Grundlagen und apparative Umsetzung.* Springer Berlin, Heidelberg, 2013.

[86] J.R. Grace, T.M. Knowlton and A.A. Avidan. *Circulating Fluidized-Beds.* Chapman & Hall, London (UK), 1997.

[87] P. Basu. *Combustion and Gasification in Fluidized Beds.* Taylor & Francis Inc, 1st Edition. New York (USA), 2006.

[88] J. Zelkowski. Kohlecharakterisierung und Kohleverbrennung, Kohle als Brennstoff, Physik und Theorie der Kohleverbrennung, Technik. VGB PowerTech Services, 2004.

[89]   F.D. Hernandez-Atonal, C. Ryu, V.N. Sharifi and J. Swithenbank. Combustion of refuse-derived fuel in a fluidised bed. *Chemical Engineering Science*, 62(1-2): 627-635, 2007. DOI: 10.1016/j.ces.2006.09.025.

[90]   F. Sher, M.A. Pans, D.T. Afilaka, C. Sun and H. Liu. Experimental investigation of woody and non-woody biomass combustion in a bubbling fluidised bed combustor focusing on gaseous emissions and temperature profiles. *Energy*, 141(15): 2069-2080, 2017. DOI: 10.1016/j.energy.2017.11.118.

[91]   P. Basu. Combustion of coal in circulating fluidized-bed boilers: a review. *Chemical Engineering Science*, 54(22): 5547-5557, 1999. DOI: 10.1016/S0009-2509(99)00285-7.

[92]   H.I. Mathekga, B.O. Oboirien and B.C. North. A review of oxy-fuel combustion in fluidized bed reactors. *International Journal of Energy Research*, 40(7): 878-902, 2016. DOI: 10.1002/er.3486

[93]   M. de las Obras-Loscertales, A. Rufas, L.F. de Diego, F. Garcìa-Labiano, P. Gayàn, A. Abad and J. Adànez. Effects of temperature and flue gas recycle on the SO_2 and NO_x emissions in an oxy-fuel fluidized bed combustor. *Energy Procedia*, 37: 1275-1282, 2013. DOI: 10.1016/j.egypro.2013.06.002.

[94]   D. Fleig, E. Vainio, K. Andersson, A. Brink, F. Johnsson and M. Hupa. Evaluation of SO_3 measurement techniques in air and oxy-fuel combustion. *Energy & Fuels*, 26(9): 5537-5549, 2012. DOI: 10.1021/ef301127x.

[95]   A. Lyngfelt and B. Leckner. Sulphur capture in circulating fluidized-bed boilers: can the efficiency be predicted? *Chemical Engineering Science*, 54(22): 5573-5584, 1999. DOI: 10.1016/S0009-2509(99)00290-0.

[96]   M. Gómez, A. Fernández, I. Llavona and R. Kuivalainen. Experiences in sulphur capture in a 30 MW$_{th}$ Circulating Fluidized Bed boiler under oxy-combustion conditions. *Applied Thermal Engineering*, 65(1-2): 617-622, 2014. DOI: 10.1016/j.applthermaleng.2014.01.012.

[97]   W. Li, S. Li, Q. Ren, L. Tan, H. Li, J. Liu and Q. Lu. Study of oxy-fuel coal combustion in a 0.1 MW$_{th}$ circulating fluidized bed at high oxygen concentrations. *Energy & Fuels*, 28(2): 1249-1254, 2014. DOI: 10.1021/ef4020422.

[98]   G Piao, S. Aono, M. Kondoh, R. Yamazaki and S. Mori. Combustion test of refuse derived fuel in a fluidized bed. *Waste Management*, 20(5-6): 443-447, 2000. DOI: 10.1016/S0956-053X(00)00009-X.

[99]   E. Desroches-Ducarne, E. Marty, G. Martin, L. Delfosse and A. Nordin, Effect of operating conditions on HCl emission from municipal solid waste combustion in a laboratory-scale fluidized bed incinerator. *Environmental Engineering Science*, 15(4): 279-289, 1998. DOI: 10.1089/ees.1998.15.279.

[100] D.A. Wenz, I. Johnson and R.D. Wolson. $CaCl_2$-rich region of the $CaCl_2$-CaF_2-CaO system. *Journal of Chemical & Engineering Data*, 14(2): 250-252, 1969. DOI: 10.1021/je60041a027.

[101] E. Vainio, P. Yrjas, M. Zevenhoven, A. Brink, T. Laurèn, M. Hupa, T. Kajolinna, and H. Vesala. The fate of chlorine, sulfur, and potassium during co-combustion of bark, sludge, and solid recovered fuel in an industrial BFB boiler. *Fuel Processing Technology*, 105: 59-68, 2013. DOI: 10.1016/j.fuproc.2011.08.021.

[102] J. Partanen, P. Backman, R. Backman and M. Hupa. Absorption of HCl by limestone in hot flue gases. Part I: the effects of temperature, gas atmosphere and absorbent quality. *Fuel*, 84(12-13): 1664-1673, 2005. DOI: 10.1016/j.fuel.2005.02.011.

[103] W. Wang, Z. Ye and I. Bjerle. The kinetics of the reaction of hydrogen chloride with fresh and spent Ca-based desulfurization sorbents. *Fuel*, 75(2): 207-212, 1996. DOI: 10.1016/0016-2361(95)00242-1.

[104] Y. Li, H. Wang, L. Jiang, W. Zhang, R. Li and Y. Chi. HCl and PCDD/Fs emission characteristics from incineration of source-classified combustible solid waste in fluidized bed. *RSC Advances*, 5: 67866-67873, 2015. DOI: 10.1039/C5RA08722H.

[105] E. Ferrer, M. Aho, J. Silvennoinen and R.-V. Nurminen. Fluidized bed combustion of refuse-derived fuel in presence of protective coal ash. *Fuel Processing Technology*, 87(1): 33-44, 2005. DOI: 10.1016/j.fuproc.2005.04.004.

[106] J. Partanen, P. Backman, R. Backman, M. Hupa. Formation of calcium chloride and its interaction with the sand particles during fluidised bed combustion. *17th Fluidized Bed Combustion Conference*, May 18 - 21, Jacksonville, Florida USA, 2003.

[107] T. Shimizu, T. Hirama, H. Hosoda, K. Kitano, M. Inagaki and K. Tejima. A twin fluid-bed reactor for removal of CO_2 from combustion processes. *Chemical Engineering Research and Design*, 77(1): 62-68, 1999. DOI: 10.1205/026387699525882.

[108] D.P. Hanak, S. Michalski and V. Manovic. From post-combustion carbon capture to sorption-enhanced hydrogen production: A state-of-the-art review of carbonate looping process feasibility. *Energy Conversion and Management*, 177(1): 428-452, 2018. DOI: 10.1016/j.enconman.2018.09.058.

[109] D.P. Hanak, E.J. Anthony and V. Manovic. A review of developments in pilot-plant testing and modelling of calcium looping process for CO_2 capture from power generation systems. *Energy & Environmental Science*, 8: 2199-2249, 2015. DOI: 10.1039/C5EE01228G.

[110] C.C. Dean, J. Blamey, N.H. Florin, M.J. Al-Jeboori and P.S. Fennell. The calcium looping cycle for CO_2 capture from power generation, cement manufacture and hydrogen production. *Chemical Engineering Research and Design*, 89(6): 836-855, 2011. DOI: 10.1016/j.cherd.2010.10.013.

[111] J.C. Abanades, E.J. Anthony, J. Wang and J.E. Oakey, Fluidized bed combustion systems integrating CO_2 capture with CaO. *Environment Science Technology*, 39(8): 2861-2866, 2005. DOI: 10.1021/es0496221.

[112] F.R. Fernandez and J.C. Abanades. CO_2 capture from the calcination of $CaCO_3$ using iron oxide as heat carrier. *Journal of Cleaner Production*, 112(1): 1211-1217, 2016. DOI: 10.1016/j.jclepro.2015.06.010.

[113] I. Martinez, R. Murillo, G. Grasa, N. Rodriguez and J.C. Abanades. Conceptual design of a three fluidized beds combustion system capturing CO_2 with CaO. *International Journal of Greenhouse Gas Control*, 5(3): 498-504, 2010. DOI: 10.1016/j.ijggc.2010.04.017

[114] D. Hoeftberger and J. Karl, The indirectly heated carbonate looping process for CO_2 capture - A concept with heat pipe heat exchanger. *Journal of Energy Resources Technology*, 138(4): 042211, 2016. DOI: 10.1115/1.4033302.

[115] M. Reitz. *Experimentelle Untersuchung und Bewertung eines indirekt beheizten Carbonate-Looping-Prozesses*. Ph.D. Thesis, Technische Universität Darmstadt, 2017.

[116] F. Garcia-Labiano, A. Abad, L.F. de Diego, P. Gayán and J. Adanez. Calcination of calcium-based sorbents at pressure in a broad range of CO_2 concentrations. *Chemical Engineering Science*, 57(13): 2381-2393., 2002. DOI: 10.1016/S0009-2509(02)00137-9.

[117] E.H. Baker. The calcium oxide-carbon dioxide system in pressure range 1-300 atmospheres. *Journal of the Chemical Society*, 0: 464-470, 1962. DOI: 10.1039/JR9620000464.

[118] K.J. Hill and E.R.S. Winter. Thermal dissociation pressure of calcium carbonate. *Journal of Physics and Chemistry*, 60(10): 1361-1362, 1956. DOI: 10.1021/j150544a005.

[119] J. Johnston. The thermal dissociation of calcium carbonate. *Journal of the American Chemistry Society*, 32(8): 938-946, 1910. DOI: 10.1021/ja01926a006.

[120] D.Y. Lu, R.W. Hughes and E.J. Anthony. Ca-based sorbent looping combustion for CO_2 capture in pilot-scale dual fluidized beds. *Fuel Processing Technology*, 89(12): 1386-1395, 2008. DOI: 10.1016/j.fuproc.2008.06.011.

[121] H. Dieter, A.R. Bidwe, G. Varela-Duelli, A. Charitos, C. Hawthorne and G. Scheffknecht. Development of the calcium looping CO_2 capture technology from lab to pilot scale at IFK, University of Stuttgart. *Fuel*, 127: 23-37, 2014. DOI: 10.1016/j.fuel.2014.01.063.

[122] D. Alvarez and J.C. Abanades. Determination of the critical product layer thickness in the reaction of CaO with CO_2. *Industrial & Engineering Chemistry Research*, 44(15): 5608-5615, 2005. DOI: 10.1021/ie050305s.

[123] G.S. Grasa and J.C. Abanades. CO_2 capture capacity of CaO in long series of carbonation/calcination cycles. *Industrial & Engineering Chemistry Research*, 45(26): 8846-8851, 2006. DOI: 10.1021/ie0606946.

[124] J.M. Valverde, P.E. Sanchez-Jimenez, L.A. Perez-Maqueda. Ca-looping for postcombustion CO_2 capture: A comparative analysis on the performances of dolomite and limestone. *Applied Energy*, 138(5): 202-215, 2015. DOI: 10.1016/j.apenergy.2014.10.087.

[125] P. Sun, J.R. Grace, C.J. Lim and E.J. Anthony. Investigation of attempts to improve cyclic CO_2 capture by sorbent hydration and modification. *Industrial & Engineering Chemistry Research*, 47(6): 2024-2032, 2008. DOI: 10.1021/ie070335q.

[126] V. Manovic and E.J. Anthony. Carbonation of CaO-based sorbents enhanced by steam addition. *Industrial & Engineering Chemistry Research*, 49(19): 9105-9110, 2010. DOI: 10.1021/ie101352s.

[127] R.T. Symonds, D.Y. Lu, R.W. Hughes, E.J. Anthony and A. Macchi. CO_2 capture from simulated syngas via cyclic carbonation/calcination for a naturally occurring limestone: Pilot-plant testing. *Industrial & Engineering Chemistry Research*, 48(18): 8431-8440, 2009. DOI: 10.1021/ie900645x.

[128] F. Donat, N.H. Florin, E.J. Anthony and P.S. Fennell. Influence of high-temperature steam on the reactivity of CaO sorbent for CO_2 capture. *Environmental Science & Technology*, 46(2): 1262-1269, 2011. DOI: 10.1021/es202679w.

[129] J. Blamey, E.J. Anthony, J. Wang and P.S. Fennell. The calcium looping cycle for large-scale CO_2 capture. *Progress in Energy and Combustion Science*, 36(2): 206-279, 2010. DOI: 10.1016/j.pecs.2009.10.001.

[130] S. Champagne, D.Y. Lu, R.T. Symonds, A. Macchi and E.J. Anthony. The effect of steam addition to the calciner in a calcium looping pilot plant. *Powder Technology*, 290: 114-123, 2016. DOI: 10.1016/j.powtec.2015.07.039.

[131] N. Rodríguez, M. Alonso, J.C. Abanades. Average activity of CaO particles in a calcium looping system. *Chemical Engineering Journal*, 156(2): 388-394, 2010. DOI: 10.1016/j.cej.2009.10.055.

[132] B.R. Stanmore and P. Gilot. Review-calcination and carbonation of limestone during thermal cycling for CO_2 sequestration. *Fuel Processing Technology*, 86(16): 1707-1743, 2005. DOI: 10.1016/j.fuproc.2005.01.023.

[133] P. Sun, J.R. Grace, C.J. Lim and E.J. Anthony. The effect of CaO sintering on cyclic CO_2 capture in energy systems. *Environmental and Energy Engineering*, 53(9): 2432-2442, 2007. DOI: 10.1002/aic.11251.

[134] J. Blamey, P.S. Fennell and D.R. Dugwell. Optimisation of regeneration strategies for exhausted sorbents for CO_2. *Proceedings of the 7th European conference on coal research and its applications, coal research forum*, Cardiff (UK), 2008.

[135] R.H. Borgwardt. Calcium oxide sintering in atmospheres containing water and carbon dioxide. *Industrial & Engineering Chemistry Research*, 28(4): 493-500, 1989. DOI: 10.1021/ie00088a019.

[136] E.J. Anthony and D.L. Granatstein. Sulfation phenomena in fluidized bed combustion systems. *Progress in Energy and Combustion Science*, 27(2): 215-236, 2001. DOI: 10.1016/S0360-1285(00)00021-6.

[137] R.T. Symonds, D.Y. Lu, A. Macchi, R.W. Hughes, and E.J. Anthony. The effect of HCl and steam on cyclic CO_2 capture performance in calcium looping systems. *Chemical Engineering Science*, (Article in press), 2017. DOI: 10.1016/j.ces.2017.08.019.

[138] Y. Li, X. Ma, W. Wang, C. Chi, J. Shi and L. Duan. Enhanced CO_2 capture capacity of limestone by discontinuous addition of hydrogen chloride in carbonation at calcium looping

conditions. *Chemical Engineering Journal*, 316: 438-448, 2017. DOI: 10.1016/j.cej.2017.01.127.

[139] H.F.W. Taylor. *Cement chemistry*. 2nd Edition, Thomas Telford Ltd, 1997.

[140] F. Scala. *Fluidized Bed Technologies for Near-Zero Emission Combustion and Gasification*. Woodhead Publishing Series in Energy, 2013.

[141] C.R. Bemrose and J. Bridgwater. A review of attrition and attrition test methods. *Powder Technology*, 49(2): 97-126, 1987. DOI: 10.1016/0032-5910(87)80054-2.

[142] A. Di Benedetto and P. Salatino. Modelling attrition of limestone during calcination and sulfation in a fluidized bed reactor. *Powder Technology*, 95(2): 119-128, 1998. DOI: 10.1016/S0032-5910(97)03327-5.

[143] B. Schüppel. *Impact of improved sorbents on the performance of the carbonate looping process*. Ph.D. Thesis, Technische Universität Darmstadt, 2012.

[144] F. Montagnaro, P. Salatino and F. Scala. The influence of temperature on limestone sulfation and attrition under fluidized bed combustion conditions. *Experimental Thermal and Fluid Science*, 34(3): 352-358, 2010. DOI: 10.1016/j.expthermflusci.2009.10.013.

[145] J. Saastamoinen, T. Pikkarainen, A. Tourunen, M. Räsänen and T. Jäntti. Model of fragmentation of limestone particles during thermal shock and calcination in fluidized beds. *Powder Technology*, 187(3): 244-251, 2008. DOI: 10.1016/j.powtec.2008.02.016.

[146] L. Jia, R. Hughes, D. Lu, E.J. Anthony and I. Lau. Attrition of calcining limestones in circulating fluidized-bed systems. *Industrial & Engineering Chemistry Research*, 46(15): 5199-5209, 2007. DOI: 10.1021/ie061212t.

[147] A. Charitos, C. Hawthorne, A.R. Bidwe, S. Sivalingam, A. Schuster, H. Spliethoff and G. Scheffknecht. Parametric investigation of the calcium looping process for CO_2 capture in a 10 kW_{th} dual fluidized bed. *International Journal of Greenhouse Gas Control*, 4(5): 776-784, 2010. DOI: 10.1016/j.ijggc.2010.04.009.

[148] G. Duelli (Varela), A. Charitos, M.E. Diego, E. Stavroulakis, H. Dieter and G. Scheffknecht. Investigations at a 10 kW_{th} calcium looping dual fluidized bed facility: Limestone calcination and CO_2 capture under high CO_2 and water vapor atmosphere. *International Journal of Greenhouse Gas Control*, 33: 103-112, 2015. DOI: 10.1016/j.ijggc.2014.12.006.

[149] F. Fan, L. Z.-S. Li and N.-s. Cai. Continuous CO_2 capture from flue gases using a dual fluidized bed reactor with calcium-based sorbent. *Industrial & Engineering Chemistry Research*, 48(24): 11140-11147, 2009. DOI: 10.1021/ie901128r.

[150] A. Cotton, K.N. Finney, K. Patchigolla, R.E.A. Eatwell-Hall, J.E. Oakey, J. Swithenbank and V. Sharifi. Quantification of trace elements emission from low-carbon emission energy sources: (I) Ca-looping cycle for post-combustion CO_2 capture and (II) fixed bed, air blown down-draft gasifier. *Chemical Engineering Science*, 107: 13-29, 2014. DOI: 10.1016/j.ces.2013.11.035.

[151] B. Gonzàlez, M. Alonso and J.C. Abanades, Sorbent attrition in a carbonation/calcination pilot plant for capturing CO_2 from flue gases. *Fuel*, 89(10): 2918-2924, 2010. DOI: 10.1016/j.fuel.2010.01.019.

[152] A. Charitos, N. Rodriguez, C. Hawthorne, M. Alonso, M. Zieba, B. Arias, G. Kopanakis, G. Scheffknecht and J.C. Abanades. Experimental validation of the calcium looping CO_2 capture process with two circulating fluidized bed carbonator reactors. *Industrial & Engineering Chemistry Research*, 50(16): 9685-9695, 2011. DOI: 10.1021/ie200579f.

[153] R.W. Hughes, D.Y. Lu, E.J. Anthony and A. Macchi. Design, process simulation and construction of an atmospheric dual fluidized bed combustion system for in situ CO_2 capture using high-temperature sorbents. *Fuel Processing Technology*, 86(14-15): 1523-1531, 2005. DOI: 10.1016/j.fuproc.2005.01.006.

[154] W. Wang, S. Ramkumar, S. Li, D. Wong, M. Iyer, B. Sakadjian, R.M. Statnick and L.-S. Fan. Subpilot demonstration of the carbonation-calcination reaction (CCR) process: High-

temperature CO_2 and sulfur capture from coal-fired power plants. *Industrial & Engineering Chemistry Research*, 49(11): 5094-5101, 2010. DOI: 10.1021/ie901509k.

[155]   M. Alonso, M.E. Diego, C. Pèrez, J.R. Chamberlain and J.C. Abanades. Biomass combustion with in situ CO_2 capture by CaO in a 300 kWth circulating fluidized bed facility. *International Journal of Greenhouse Gas Control*, 29: 142-152, 2014. DOI: 10.1016/j.ijggc.2014.08.002.

[156]   M.E. Diego and M. Alonso, Operational feasibility of biomass combustion with in situ CO_2 capture by CaO during 360 h in a 300 kW_{th} calcium looping facility. *Fuel*, 181: 325-329, 2016. DOI: 10.1016/j.fuel.2016.04.128.

[157]   M. Reitz, M. Junk, J. Ströhle and B. Epple. Design and operation of a 300 kW_{th} indirectly heated carbonate looping pilot plant. *International Journal of Greenhouse Gas Control*, 54(1): 272-281, 2016. DOI: 10.10.1016/j.ijggc.2016.09.016.

[158]   M. Junk, M. Reitz, J. Ströhle and B. Epple. Thermodynamic evaluation and cold flow model testing of an indirectly heated carbonate looping process. *Chemical Engineering & Technology*, 36(9): 1479-1487, 2013. DOI: 10.1002/ceat.201300019.

[159]   A. Galloy, J. Ströhle and B. Epple. Post-combustion CO_2 capture experiments in a 1 MW_{th} carbonate looping pilot. *VGB PowerTech*, 6: 33-37, 2012.

[160]   M. Helbig, J. Hilz, M. Haaf, A. Daikeler, J. Ströhle and B. Epple. Long-term carbonate looping testing in a 1 MW_{th} pilot plant with hard coal and lignite. *Energy Procedia*, 114: 179-190, 2017. DOI: 10.1016/j.egypro.2017.03.1160.

[161]   M.E. Diego, B. Arias and J.C. Abanades. Evolution of the CO_2 carrying capacity of CaO particles in a large calcium looping pilot plant. *International Journal of Greenhouse Gas Control*, 62: 69-75, 2017. DOI: 10.1016/j.ijggc.2017.04.005.

[162]   B. Arias, M.E. Diego, J.C. Abanades, M. Lorenzo, L. Diaz, D. Martinez, J. Alvarez and A. Sanchez-Biezma. Demonstration of steady state CO_2 capture in a 1.7 MW_{th} calcium looping pilot. *International Journal of Greenhouse Gas Control*, 18:237-245, 2013. DOI: 10.1016/j.ijggc.2013.07.014.

[163]   M.-H. Chang, C.-M. Huang, W.-H. Liu, W.-C. Chen, J.-Y. Cheng, W. Chen, T.-W. Wen, S. Ouyang, C.-H. Shen and H.-W. Hsu. Design and experimental investigation of the calcium looping process for 3-kW_{th} and 1.9-MW_{th} facilities. *Chemical Engineering & Technology*, 36(9): 1525-1532, 2013. DOI: 10.1002/ceat.201300081.

[164]   M.-H. Chang, W.-C. Chen, C.-M. Huang, W.-H. Liu, Y.-C. Chou, W.-C. Chang, W. Chen, J.-Y. Cheng, K.-E. Huang and H.-W. Hsu, Design and experimental testing of a 1.9 MW_{th} calcium looping pilot plant. *Energy Procedia*, 63: 2100-2108, 2014. DOI: 10.1016/j.egypro.2014.11.226.

[165]   P. Ohlemüller, M. Reitz, J. Ströhle and B. Epple. Investigation of chemical looping combustion of natural at 1 MW_{th} scale. *Proceedings of the Combustion Institute*, 37(4): 4353-4360, 2019. DOI: 10.1016/j.proci.2018.07.035.

[166]   P. Ohlemüller, J. Ströhle and B. Epple. Chemical looping combustion of hard coal and torrefied biomass in a 1 MW_{th} pilot plant. *International Journal of Greenhouse Gas Control*, 65:149-159, 2017. DOI: 10.1016/j.ijggc.2017.08.013.

[167]   D. Krause, P. Herdel, J. Ströhle and B. Epple. HTW gasification of high volatile bituminous coal in a 500 kW_{th} pilot plant. *Fuel*, 250: 306-314, 2019. DOI: 10.1016/j.fuel.2019.04.014.

[168]   P. Herdel, D. Krause, J. Peters, B. Kolmorgen, J. Ströhle and B. Epple. Experimental investigations in a demonstration plant for fluidized bed gasification of multiple feedstock`s in 0.5 MW_{th} scale. *Fuel*, 205:286-296, 2017. DOI: 10.1016/j.fuel.2017.05.058.

[169]   M. Carvalho, J. Peters, D. Krause, J. Ströhle, B. Epple, A. Rokka, V. Barisic and T. Eriksson. Towards optimization of multi-pollutant emissions control - CFBS pilot tests in 1 MW_{th} scale. Proceedings of the 23rd International Conference on FBC, Seoul (South Korea): A6-2: 217-225, 2018.

[170] J. Kremer, A. Galloy, J. Ströhle and B. Epple. Continuous CO_2 capture in a 1-MW$_{th}$ carbonate looping pilot plant. *Chemical Engineering & Technology*, 36(9): 1518-1524, 2013. DOI: 10.1002/ceat.201300084.

[171] J. Ströhle, M. Junk, J. Kremer, A. Galloy and B. Epple. Carbonate looping experiments in a 1 MW$_{th}$ pilot plant and model validation. *Fuel*, 127: 13-22, 2014. DOI: 10.1016/j.fuel.2013.12.043.

[172] SCARLET, 2020. *Scale-up of Calcium Carbonate Looping Technology for efficient CO_2 Capture from Power and Industrial Plants*. URL: http://www.project-scarlet.eu/wordpress/. Cited: 09.04.2020.

[173] J. Hilz, M. Helbig, M. Haaf, A. Daikeler, J. Ströhle and B. Epple. Investigation of the fuel influence on the carbonate looping process in 1 MW$_{th}$ scale. *Fuel Processing Technology*, 169: 170-177, 2018. DOI: 10.1016/j.fuproc.2017.09.016.

[174] J. Hilz, M. Helbig, M. Haaf, A. Daikeler, J. Ströhle and B. Epple. Long-term pilot testing of the carbonate looping process in 1 MW$_{th}$ scale. *Fuel*, 210: 892-899, 2017. DOI: 10.1016/j.fuel.2017.08.105.

[175] M.E. Diego, B. Arias, A. Mèndez, M. Lorenzo, L. Diaz, A. Sanchez-Biezma and J.C. Abanades. Experimental testing of a sorbent reactivity process in La Pereda 1.7 MW$_{th}$ calcium looping pilot plant. *International Journal of Greenhouse Gas Control*, 50: 14-22, 2016. DOI: 10.1016/j.ijggc.2016.04.008.

[176] Arias, B., M.E. Diego, A. Mèndez, M. Alonso and J.C. Abanades. Calcium looping performance under extreme oxy-fuel combustion conditions in the calciner. *Fuel*, 222: 711-717, 2018. DOI: 10.1016/j.fuel.2018.02.163.

[177] M.E. Diego, B. Arias and J.C. Abanades. Investigation of the dynamic evolution of the CO_2 carrying capacity of solids with time in La Pereda 1.7 MW$_{th}$ calcium looping pilot plant. *International Journal of Greenhouse Gas Control*, 92: 102856, 2020. DOI: 10.1016/j.ijggc.2019.102856.

[178] J.C. Abanades and D. Alvarez. Conversion Limits in the Reaction of CO_2 with Lime. *Energy & Fuels*, 17(2): 308-315, 2003. DOI: 10.1021/ef020152a.

[179] L. Zhen-shan, C. Ning-sheng and E. Croiset. Process analysis of CO_2 capture from flue gas using carbonation/calcination cycles. *Environmental and Energy Engineering*, 54(7): 1912-1925, 2008. DOI: 10.1002/aic.11486.

[180] M.C. Romano. Modeling the carbonator of a Ca-looping process for CO_2 capture from power plant flue gas. *Chemical Engineering Science*, 69(1): 257-269, 2012. DOI: 10.1016/j.ces.2011.10.041.

[181] G.S. Grasa, M. Alonso and J.C. Abanades, Sulfation of CaO particles in a carbonation/calcination loop to capture CO_2. *Industrial & Engineering Chemistry Research*, 47(5): 1630-1635, 2008. DOI: 10.1021/ie070937+.

[182] J.C. Abanades. The maximum capture efficiency of CO_2 using a carbonation/calcination cycle of $CaO/CaCO_3$. *Chemical Engineering Journal*, 90(3): 303-306, 2002. DOI: 10.1016/S1385-8947(02)00126-2.

[183] I. Martinez, G. Grasa, J. Parkkinen, T. Tynjälä, T. Hyppänen, R. Murillo and M.C. Romano. Review and research needs of Ca-looping systems modelling for post-combustion CO_2 capture applications. *International Journal of Greenhouse Gas Control*, 50: 271-304, 2016. DOI: 10.1016/j.ijggc.2016.04.002.

[184] J.C. Abanades, E.J. Anthony, D.Y. Lu, C. Salvador and D. Alvarez. Capture of CO_2 from combustion gases in a fluidized bed of CaO. *AlChE Journal*, 50(7): 1614-1622, 2004. DOI: 10.1002/aic.10132.

[185] S.K. Bhatia and D.D. Perlmutter. Effect of the product layer on the kinetics of the CO_2-lime reaction. *AlChE Journal*, 29(1): 79-86, 1983. DOI: 10.1002/aic.690290111.

[186] C. Hawthorne, A. Charitos, C.A. Perez-Pulido, Z. Bing and G. Scheffknect. Design of a dual fluidized bed system for the post-combustion removal of CO_2 using CaO, part I: CFB

carbonator model. *Proceedings of the 9th International Conference on Circulating Fluidized Beds*, 2008.

[187]   T. S. Pugsley, F. Berruti. A predictive hydrodynamic model for circulating fluidized bed risers. *Powder Technology*, 89(1): 57-69, 1996. DOI: 10.1016/S0032-5910(96)03154-3.

[188]   A. Lasheras, J. Ströhle, A. Galloy and B. Epple. Carbonate looping process simulation using a 1D fluidized bed model for the carbonator. *International Journal of Greenhouse Gas Control*, 5(4): 686-693, 2011. DOI: 10.1016/j.ijggc.2011.01.005.

[189]   D. Kunii and O. Levenspiel. Circulating fluidized-bed reactors. *Chemical Engineering Science*, 52(15): 2471-2482, 1997. DOI: 10.1016/S0009-2509(97)00066-3.

[190]   M. Haaf, A. Stroh, J. Hilz, M. Helbig, J. Ströhle and B. Epple. Process modelling of the calcium looping process and validation against 1 MW$_{th}$ pilot testing. *Energy Procedia*, 114: 167-178, 2017. DOI: 10.1016/j.egypro.2017.03.1159.

[191]   J. Hilz, M. Haaf, M. Helbig, N. Lindqvist, J. Ströhle and B. Epple. Scale-up of the carbonate looping process to a 20 MW$_{th}$ pilot plant based on long-term pilot tests. *International Journal of Greenhouse Gas Control*, 88: 332-341, 2019. DOI: 10.1016/j.ijggc.2019.04.026.

[192]   M. Haaf, J. Hilz, M. Helbig, C. Weingärtner, O. Stallmann, J. Ströhle and B. Epple. Assessment of the operability of a 20 MW$_{th}$ calcium looping demonstration plant by advanced process modelling. *International Journal of Greenhouse Gas Control*, 75: 224-234, 2018. DOI: 10.1016/j.ijggc.2018.05.014.

[193]   J. Ylätalo, J. Ritvanen, B. Arias, T. Tynjälä and T. Hyppänen. 1-Dimensional modelling and simulation of the calcium looping process. *International Journal of Greenhouse Gas Control*, 9: 130-135, 2012. DOI: 10.1016/j.ijggc.2012.03.008.

[194]   F. Johnsson and B. Leckner. Vertical distribution of solid in a CFB furnace. *Proceedings of the 13th International Conference of Fluidized Bed Combustion*, 1995.

[195]   A.-M. Cormos and A. Simon. Assessment of CO_2 capture by calcium looping (CaL) process in a flexible power plant operation scenario. *Applied Thermal Engineering*, 80: 319-327, 2015. DOI: 10.1016/j.applthermaleng.2015.01.059.

[196]   A.-M. Cormos and A. Simon. Dynamic modelling of CO_2 capture by calcium-looping cycle. *Chemical Engineering Transactions*, 35: 421-426, 2013. DOI: 10.3303/CET1335070.

[197]   K. Atsonios, M. Zeneli, A. Nikolopoulos, N. Nikolopoulos, P. Grammelis and E. Kakaras. Calcium looping process simulation based on an advanced thermodynamic model combined with CFD analysis. *Fuel*, 153(1): 370-381, 2015. DOI: 10.1016/j.fuel.2015.03.014.

[198]   C. Hawthorne, M. Trossmann, P.G. Cifre, A. Schuster and G. Scheffknecht. Simulation of the carbonate looping power cycle. *Energy Procedia*, 1: 1387-1394, 2009. DOI: 10.1016/j.egypro.2009.01.182.

[199]   Y. Lara, P. Lisbona, A. Martinez and L.M. Romeo. A systematic approach for high temperature looping cycles integration. *Fuel*, 127: 4-12, 2013. DOI: 10.1016/j.fuel.2013.09.062.

[200]   J. Ströhle, A. Lasheras, A. Galloy and B. Epple. Simulation of the carbonate looping process for post-combustion CO_2 capture from a coal-fired power plant. *Chemical Engineering & Technology*, 32(3): 435-442, 2009. DOI: 10.1002/ceat.200800569.

[201]   I. Martinez, R. Murillo, G. Grasa, J.C. Abanades. Integration of a Ca looping system for CO_2 capture in existing power plants. *AIChE Journal*, 57(9): 2599-2607, 2011. DOI: 10.1002/aic.12461.

[202]   I. Vorrias, K. Atsonios, A. Nikolopoulos, N. Nikolopoulos, P. Grammelis and E. Kakaras. Calcium looping for CO_2 capture from a lignite fired power plant. *Fuel*, 113: 826-836, 2013. DOI: 10.1016/j.fuel.2012.12.087.

[203]   A. Rolfe, Y. Huang, M. Haaf, S. Rezvani, D. Mcliveen-Wright and N.J. Hewitt. Integration of the calcium carbonate looping process into an existing pulverized coal-fired power plant for CO_2 capture: Techno-economic and environmental evaluation. *Applied Energy*, 222: 169-179, 2018. DOI: 10.1016/j.apenergy.2018.03.160.

[204] E.S. Rubin, J.E. Davison and H.J. Herzog. The cost of CO_2 capture and storage. *International Journal of Greenhouse Gas Control*, 40: 378-400, 2015. DOI: 10.1016/j.ijggc.2015.05.018

[205] M. Zhao, A.I. Minett and A.T. Harris. A review of techno-economic models for the retrofitting of conventional pulverized-coal power plants for post-combustion capture (PCC) of CO_2. *Energy & Environmental Science*, 6: 25-40, 2013. DOI: 10.1039/C2EE22890D.

[206] A. Rolfe, Y. Huang, M. Haaf, S. Rezvani, A. Dave and N.J. Hewitt. Techno-economic and environmental analysis of calcium carbonate looping for CO_2 capture from a pulverised coal-fired power plant. *Energy Procedia*, 142: 3447-3453, 2017. DOI: 10.1016/j.egypro.2017.12.228

[207] Y. Yongping, Z. Rongrong, D. Liqiang, M. Kavosh, K. Patchigolla and J. Oakey. Integration and evaluation of a power plant with a CaO-based CO_2 capture system. *International Journal of Greenhouse Gas Control*, 4(4): 603-612, 2010. DOI: 10.1016/j.ijggc.2010.01.004.

[208] J. Kotowicz, S. Michalski and M. Brzeczek. The characteristics of a modern oxy-fuel power plant. Energies, 12(7): 3374, 2019. DOI: 10.3390/en12173374.

[209] C.-C. Cormos. Oxy-combustion of coal, lignite and biomass: A techno-economic analysis for a large scale carbon capture and storage (CCS) project in Romania. *Fuel*, 169: 50-57, 2016. DOI: 10.1016/j.fuel.2015.12.005.

[210] M. R.M. Abu-Zahra, L.H.J. Schneiders, J.P.M. Niederer, P.H.M. Feron and G.F. Versteeg. CO_2 capture from power plants: Part I. A parametric study of the technical performance based on monoethanolamine. *International Journal of Greenhouse Gas Control*, 1(1): 37-46, 2007. DOI: 10.1016/S1750-5836(06)00007-7.

[211] International Energy Agency Greenhouse Gas R&D Program (IEAGHG). *CO2 Capture at Coal Based Power and Hydrogen Plants*. 2014/03. 2014.

[212] P. Maas, N. Nauels, L. Zhao, P. Markewitz, V. Scherer, M. Modigell, D. Stolten and J.-F. Hake. Energetic and economic evaluation of membrane-based carbon capture routes for power plant processes. *International Journal of Greenhouse Gas Control*, 44: 124-139, 2016. DOI: 10.1016/j.ijggc.2015.11.018.

[213] H. Zhai and E.S. Rubin. Techno-economic assessment of polymer membrane systems for postcombustion carbon capture at coal-fired power plants. *Environmental Science & Technology*, 47(6): 3006-3014, 2013. DOI: 10.1021/es3050604.

[214] S. Roussanaly, M. Vitvarova, R. Anantharaman, D. Berstad, B. Hagen, J. Jakobsen, V. Novotny and G. Skaugen. Techno-economic comparison of three technologies for pre-combustion CO_2 capture from a lignite-fired IGCC. *Frontiers of Chemical Science and Engineering*, 2019. DOI: 10.1007/s11705-019-1870-8.

[215] G.E. Klinzing, F. Rizk, R. Marcus and L.S. Leung. Pneumatic Conveying of Solids, 3rd Edition, Springer, Dordrecht Heidelberg London New York, 2010.

[216] J. Hilz. *Experimental investigation of a semi-industrial carbonate looping process for scale-up*. Ph.D. Thesis, Technische Universität Darmstadt, 2019.

[217] M. Helbig. *Experimentelle Untersuchung des Langzeitverhaltens des Carbonate-Looping-Verfahrens im 1 Megawatt-Technikum*. Ph.D. Thesis, Technische Universität Darmstadt, 2019.

[218] SUEZ Deutschland GmbH. *Ersatzbrennstoffherstellung*, 2019. URL: https://www.suez-deutschland.de/leistungen/verwertungstechnologien/. Cited: 14.01.2019.

[219] RWE Power AG. *Braunkohlestaub, Analysewerte ab Lieferwerk*. 2013.

[220] e-netz Südhessen. *Brennwerte und Gasanalyse*, 2019. URL: https://www.e-netz-suedhessen.de/privatkunden/brennwerte-und-gasanalyse/. Cited: 14.01.2019.

[221] D.A. Skoog, F.J. Holler, S.R. Crouch. Instrumentelle Analytik Grundlagen - Geräte - Anwendung, 2nd Edition, Springer Spektrum, 2013

[222] Lhoist, Rheinkalk GmbH. SOP 9.10.2070 Messung von Chlorid in Feststoffen (inkl.Sekundärbrennstoffe) DE_60_5211_LGE, 2016.

[223] DIN 66165-1. *Partikelgrößenanalyse - Siebanalyse - Teil 1: Grundlagen*. 2016.

[224] DIN 66165-2. *Partikelgrößenanalyse - Siebanalyse - Teil 2: Durchführung.* 2016

[225] M. Rydèn, P. Moldenhauer, S. Lindqvist, T. Mattisson and A. Lyngfelt. Measuring attrition resistance of oxygen carrier particles for chemical looping combustion with a customized jet cup. *Powder Technology*, 256:75-86, 2014. DOI: 10.1016/j.powtec.2014.01.085.

[226] NETZSCH-Gerätebau GmbH. *STA 449 F3 Jupiter - Technische Daten.* 2016.

[227] M. Alonso, N. Rodrìguez, G. Grasa and J.C. Abanades. Modelling of a fluidized bed carbonator reactor to capture CO_2 from a combustion flue gas. *Chemical Engineering Science*, 64(5): 883-891, 2009. DOI: 10.1016/j.ces.2008.10.044.

[228] J.C. Abanades, M. Alonso and N. Rodriguez. Experimental validation of in situ CO_2 capture with CaO during the low temperature combustion of biomass in a fluidized bed reactor. *International Journal of Greenhouse Gas Control*, 5(3): 512-520, 2011. DOI: 10.1016/j.ijggc.2010.01.006.

[229] M. Haaf, J. Hilz, J. Peters, A. Unger, J. Ströhle and B. Epple. Operation of a 1 MW_{th} Calcium Looping Pilot Plant Firing Waste-Derived Fuels in the Calciner. *Powder Technology*, 372:267-274, 2020. DOI: https://doi.org/10.1016/j.powtec.2020.05.074.

[230] L. Duan, Y. Duan, Y. Sarbassov, Y. Li, E.J. Anthony. SO_3 formation under oxy-CFB combustion conditions. *International Journal of Greenhouse Gas Control*, 43: 172-178, 2015. DOI: 10.1016/j.ijggc.2015.10.028.

[231] B. Leckner and L.-E. Amand. Emissions from a circulating and a stationary fluidized bed boiler: a comparison. *Proceedings of the 9th International Conference on Fluidized Bed Combustion*, 2:891-897, 1987.

[232] E.P. O'Neil, N.H. Ulerich, R.A. Newby and D.L. Keairns. *Criteria for the selection of SO2 sorbents for atmospheric pressure fluidized-bed combustors.* Electric Power Research Institute, Report, FP-1307, vol. 1.

[233] K. Myöhänen, T. Hyppänen, T. Pikkarainen, T. Eriksson and A. Hotta. Near zero CO_2 emissions in coal firing with oxy-fuel circulating fluidized bed boiler. *Chemical Engineering & Technology*, 32(3): 355-363, 2009. DOI: 10.1002/ceat.200800566.

[234] D. Bankiewicz, P. Vainikka, D. Lindberg, A. Frantsi, J. Silvennoinen, P. Yrjas and M. Hupa. High temperature corrosion of boiler waterwalls induced by chlorides and bromides - Part 2: Lab-scale corrosion tests and thermodynamic equilibrium modeling of ash and gaseous species. *Fuel*, 94: 240-250, 2012. DOI: 10.1016/j.fuel.2011.12.023.

[235] P. Vainikka, D. Bankiewicz, A. Frantsi, J. Silvennoinen, J. Hannula, P. Yrjas and M. Hupa. High temperature corrosion of boiler waterwalls induced by chlorides and bromides. Part 1: Occurrence of the corrosive ash forming elements in a fluidized bed boiler co-firing solid recovered fuel. *Fuel*, 90(5): 2055-2063, 2011. DOI: 10.1016/j.fuel.2011.01.020

[236] M. Haaf, J. Peters, J. Hilz, A. Unger, J. Ströhle and B. Epple. Combustion of solid recovered fuels within the calcium looping process - experimental demonstration at 1 MW_{th} scale. *Experimental Thermal and Fluid Science*, 113(1): 110023, 2020. DOI: 10.1016/j.expthermflusci.2019.110023.

[237] Z.-S. Liu, M.-Y. Wey, C.-L. Lin. Reaction characteristics of $Ca(OH)_2$, HCl and SO_2 at low temperature in a spray dryer integrated with a fabric filter. *Journal of Hazardous Materials*, 95(3): 291-304, 2002. DOI: 10.1016/S0304-3894(02)00142-5.

[238] D. Kunii and O. Levenspiel. The K-L reactor model for circulating fluidized beds. *Chemical Engineering Science*, 55(20): 4563-4570, 2000. DOI: 10.1016/S0009-2509(00)00073-7.

[239] D. Geldart, J. Cullinan, S. Georghiades, D. Gilvray and D.J. Pope. Effect of fines on entrainment from gas fluidized beds. *Transaction of the Institute of Chemical Engineers*, 57(4):269-275, 1979.

[240] P. Deuflhard and A. *Hohmann, Numerische Mathematik I, eine algorithmisch orientierte Einführung.* De Gruyter, 2008.

[241] M. Junk. *Technical and economical assessment of various carbonate looping process configurations.* Ph.D. Thesis, Technische Universität Darmstadt, 2016.

[242] Aspen Plus. *Aspen Plus User Guide*, Version 10.2. Aspen Technology, Inc, 2000.

[243] M. Obermoser, J. Fellner and H. Rechberger. Determination of reliable CO_2 emission factors for waste-to-energy plants. *Waste Management & Research*, 27(9): 907-913, 2009. DOI: 10.1177/0734242X09349763

[244] R. Anantharaman, D. Berstad, G. Cinti, E.D. Lena, M. Ganti, M. Gazzani, H. Hoppe, I. Martinez, J.G.M.-S. Monteiro, M. Romano, S. Roussanaly, E. Schols, M. Spinelli, S. Storset, P. van Os and M. Voldsund. *CEMCAP Deliverable 3.2: CEMCAP framework for comparative techno-economic analysis of CO_2 capture from cement plants*, 2017.

[245] S. Consonni and F. Viganò. Waste gasification vs. conventional waste-to-energy: A comparative evaluation of two commercial technologies. *Waste Management*, 32(4): 653-666, 2012. DOI: 10.1016/j.wasman.2011.12.019.

[246] T. Keller. La valorisation énergétique des déchets par incineration, 2011. URL: http://www.vernimmen.net/ftp/La_valorisation_energetique_des_dechets_par_incineration.p df. Cited 14.12.2019.

[247] Agence de l'Enviroment et de la Maìtrise de l'Energy (ADEME). Enquete sur les prix de l'incineration des dechtes municipaux, 2011. URL: https://www.ademe.fr/sites/default/files/assets/documents/82444_etude_incineration.pdf Cited: 14.12.2019.

[248] S.O. Gardarsdottir, E. de Lena, M. Romano, S. Roussanaly, M. Voldsund, J.-F. Perez-Calvo, D. Berstad, C. Fu, R. Anantharaman, D. Sutter, M. Gazzani, M. Mazzotti and G. Cinti, Comparison of technologies for CO_2 capture from cement production - part 2: Cost analysis. *Energies*, 12(3): 542, 2019. DOI: 10.3390/en12030542.

[249] R. Anantharaman, O. Bolland, N. Booth, E.v. Dorst, C. Ekstrom, E.S. Fernandes, F. Franco, E. Macchi, G. Manzolini, D. Nikolic, A. Pfeffer, M. Prins, S. Rezvani and L. Robinson, *DECARBit Deliverable 1.4.3: European best practice guidelines for assessment of CO_2 capture technologies*, 2011.

[250] International Energy Agency Greenhouse Gas R&D Program (IEAGHG). *Understanding the cost of retrofitting CO_2 capture in an Integrated Oil Refinery*. 2017.

[251] E.S. Rubin. Understanding the pitfalls of CCS cost estimates. *International Journal of Greenhouse Gas Control*, 10: 181-190, 2012. DOI: 10.1016/j.ijggc.2012.06.004.

[252] S. Roussanaly, J.A. Ouassou, R. Anantharaman and M. Haaf. Impact of uncertainties on the design and cost of CCS from a waste-to-energy plant. *Frontiers in Energy Research*, 8(17), 2020. DOI: 10.3389/fenrg.2020.00017.

[253] L.M. Romeo, Y. Lara, P. Lisbona and J.M. Escosa. Optimizing make-up flow in a CO_2 capture system using CaO. *Chemical Engineering Journal*, 147(2-3): 252-258, 2009. DOI: 10.1016/j.cej.2008.07.010.

[254] R. Baehr, F.L. Blangetti, B. Braun, W. Casper, H. Haase, C.H. Haefele, H. Huschauer, M. Jung and R. Krane. *Konzeption und Aufbau von Dampfkraftwerken*. Technischer Verlag Resch Koeln TÜV Rheinland, Gräfeling, 1985.

[255] R. W. Bryers. Fireside slagging, fouling, and high-temperature corrosion of heat-transfer surface due to impurities in steam-raising fuels. *Progress in Energy and Combustion Science*, 22(1): 29-120, 1996. DOI: 10.1016/0360-1285(95)00012-7.

[256] Y.-N. Chang and F.-I. Wei. High-temperature chlorine corrosion of metals and alloys. *Journal of Material Science*, 26: 3693-3698, 1991. DOI: 10.1007/BF01184958.

[257] VGB PowerTech e.V. *VGB-Standard: Maßnahmen zur Verminderung von Abzehrungen in den abfall- und biomassegefeuerten Dampferzeugern*. VGB PowerTech Service GmbH, Essen, 2018.

[258] Kümmel, J., Dampfkessel in Hausmüll- bzw. Restmüll-Verbrennungsanlagen, in Feuerungs-, Verbrennungs-, Vergasungstechniken, VDi-Bildungswerk, Editor. 1994: Düsseldorf.

[259] C. Fu, R. Anantharaman, K. Jordal and T. Gundersen, Thermal efficiency of coal-fired power plants: From theoretical to practical assessments. *Energy Conversion and Management*, 105:530-544, 2015. DOI: 10.1016/j.enconman.2015.08.019.

[260] R. Schu and R. Leithner. Waste to Energy - Higher Efficiency with External superheating. *Proceedings of the Second International Symposium on Energy from Biomass and Waste*, Venice (Italy), 2008.

[261] M. Becidan and R. Anantharaman. Dual-fuel cycles to increase the efficiency of WtE installations. *Chemical Engineering Transactions*, 29: 727-732, 2012. DOI: 10.3303/CET1229122.

[262] CO_2 European Emission Allowance, 2019, URL: https://www.finanzen.net/rohstoffe/co2-emissionsrechte. Cited: 12.12.2019,

[263] A. Bhave, R.H.S. Taylor, P. Fennell, W.R. Livingston, N. Shah, N. Mac Dowell, J. Dennis, M. Kraft, M. Pourkashanian, M. Insa, J. Jones, N. Burdett, A. Bauen, C. Beal, A. Smallbone and J. Akroyd. Screening and techno-economic assessment of biomass-based power generation with CCS technologies to meet 2050 CO_2 targets. *Applied Energy*, 190: 481-489, 2017. DOI: 10.1016/j.apenergy.2016.12.120.

[264] M. Fasihi, O. Efimova and C. Breyer. Techno-economic assessment of CO_2 direct air capture plants. *Journal of Cleaner Production*, 224: 957-980, 2019. DOI: 10.1016/j.jclepro.2019.03.086.

[265] Climeworks AG, 2019. URL: https://climeworks.shop/. Cited: 08.11.2019.

[266] United States Environmental Protection Agency (EPA). Documentation for EPA`s power sector modeling platform v6 - November 2018 Reference Case. URL: https://www.epa.gov/airmarkets/documentation-epas-power-sector-modeling-platform-v6-may-2019-reference-case. Cited 20.05.2019.

[267] J.G. Wimer and W.M. Morgan. *Quality guideline for energy system studies: cost estimation methodology for NETL assessment of power plant performance*. 2011. DOI: 10.2172/1489510.

A. Appendix

A. 1 Results of Solid Sample Analysis

Table A-1: Results of solid sample analysis for test campaign 1.

S. No.	Date	Time	Spot	CaO	CaCO$_3$	CaSO$_4$	Cl	Ash
-	dd.mm.yyyy	hh:mm	-	wt.%	wt.%	wt.%	wt.%	wt.%
1	14.08.2018	20:55	BA$_{Calc}$	67.3	20.5	2.74	0.12	9.20
2	14.08.2018	20:55	LS$_{4.1}$	74.3	12.6	0.43	0.15	12.2
3	14.08.2018	20:55	LS$_{4.4}$	87.4	0.41	0.17	0.16	11.7
4	14.08.2018	20:55	FA$_{Calc}$	51.8	36.9	0.99	2.22	9.50
29	15.08.2018	16:55	LS$_{4.1}$	73.2	12.1	0.17	0.15	14.0
30	15.08.2018	16:55	LS$_{4.4}$	86.5	0.52	0.20	0.16	12.2
67	22.08.2018	21:10	LS$_{4.1}$	69.8	11.5	0.91	0.20	16.9
68	22.08.2018	21:10	LS$_{4.4}$	82.1	0.48	0.98	0.22	15.7
74	23.08.2018	05:00	LS$_{4.1}$	71.9	11.2	1.05	0.21	15.1
75	23.08.2018	05:00	LS$_{4.4}$	82.5	0.60	1.11	0.24	15.0
98	24.08.2018	03:25	FA$_{Calc}$	53.4	24.9	1.92	3.34	18.4
99	24.08.2018	05:00	LS$_{4.1}$	72.9	9.07	1.70	0.32	15.5
100	24.08.2018	05:00	LS$_{4.4}$	80.6	1.05	1.75	0.33	15.8
101	24.08.2018	05:00	BA$_{Calc}$	76.5	4.94	1.15	0.24	16.5
102	24.08.2018	05:00	BA$_{Carb}$	53.9	7.41	1.00	0.20	35.0

Table A-2: Results of solid sample analysis for test campaign 2.

S. No.	Date	Time	Spot	CaO	CaCO$_3$	CaSO$_4$	Cl	Ash
-	dd.mm.yyyy	hh:mm	-	wt.%	wt.%	wt.%	wt.%	wt.%
15	11.09.2018	17:50	FA$_{Calc}$	42.2	42.3	1.33	2.51	12.9
17	11.09.2018	18:00	LS$_{4.1}$	29.1	58.3	0.91	0.44	11.1
18	11.09.2018	18:00	LS$_{4.4}$	36.9	49.7	1.02	0.51	11.7
19	11.09.2018	18:00	BA$_{Calc}$	53.3	28.3	0.95	0.25	16.7
22	11.09.2018	21:30	LS$_{4.1}$	38.3	45.6	1.28	0.64	14.4
23	11.09.2018	21:30	LS$_{4.4}$	46.2	34.5	1.43	0.75	17.0
50	16.09.2018	10:00	LS$_{4.1}$	65.9	20.2	0.69	0.23	12.6
51	20.09.2018	10:00	LS$_{4.4}$	71.6	16.4	0.89	0.26	10.5
81	20.09.2018	12:00	LS$_{4.1}$	70.1	12.1	0.99	0.27	16.0
82	20.09.2018	12:00	LS$_{4.4}$	64.1	16.0	1.24	0.36	17.8
83	20.09.2018	16:15	LS$_{4.1}$	55.3	28.0	0.96	0.36	15.0
84	20.09.2018	16:15	LS$_{4.4}$	63.6	18.2	1.06	0.40	16.4
87	20.09.2018	20:00	LS$_{4.1}$	53.8	52.3	1.00	0.41	10.1
88	20.09.2018	20:00	LS$_{4.4}$	39.5	48.5	1.01	0.42	10.5
99	21.09.2018	03:00	BA$_{Calc}$	30.8	53.8	0.95	0.36	13.6
100	21.09.2018	03:00	BA$_{Carb}$	38.3	42.2	1.06	0.32	17.4
101	21.09.2018	03:00	LS$_{4.1}$	63.0	21.3	1.28	0.36	13.8
102	21.09.2018	03:00	LS$_{4.4}$	81.8	2.22	1.42	0.38	13.8
104	21.09.2018	03:00	FA$_{Carb}$	53.6	30.0	1.42	1.59	14.4

A. 2 Data Plots of Carbonate Looping Test Campaigns

Within this section, the data plots for relevant process parameter of the CaL test campaigns is given. There are two (three) sets of diagrams for each test campaign. The structure for each set of figure is similar and described below.

- Figure a)
    - Left y-axis: Average reactor temperature for carbonator and calciner (T_{carb}, T_{calc})
    - Right y-axis: Mass flow of oxygen fed to calciner ($\dot{m}_{O2}$), mass flow of SRF ($\dot{m}_{SRF}$)
- Figure b)
    - Left y-axis: Total CO_2 capture rate (E_{tot}), CO_2-absorption rate (E_{carb})
    - Right y-axis: Molar flow of CO_2 absorbed ($F_{CO2,abs}$)
- Figure c)
    - Volumetric gas concentration of CO_2 at the carbonator inlet ($y_{CO2,carb,in}$) and outlet ($y_{CO2,carb,out}$), volumetric gas concentration of CO_2 and O_2 at the inlet of the calciner ($y_{CO2,calc,in}$, $y_{O2,calc,in}$), all numbers are based on dry conditions
- Figure d)
    - Left y-axis: Specific solid inventory of carbonator ($W_{s,carb}$) and calciner ($W_{s,calc}$)
    - Right y-axis: molar make-up feed (F_0)

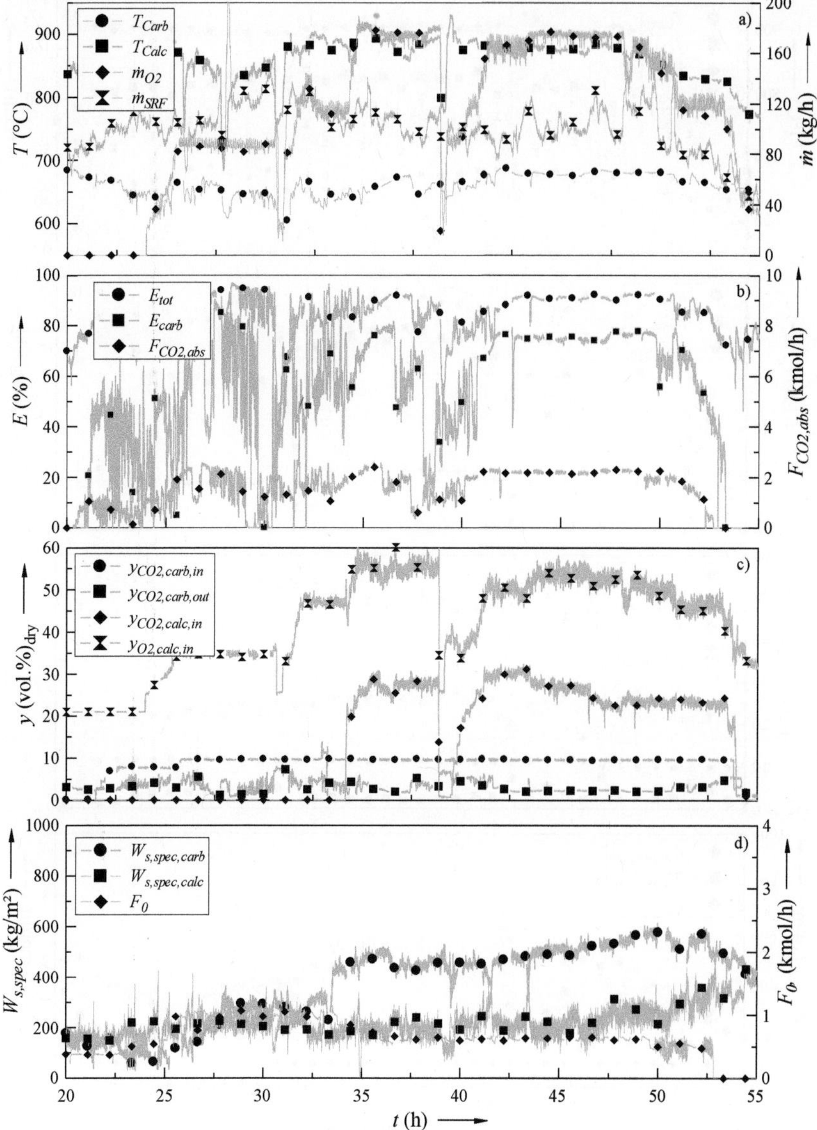

Figure A-1: Data plots (Graph a-c) of relevant process data for the first CaL test campaign (part 1).

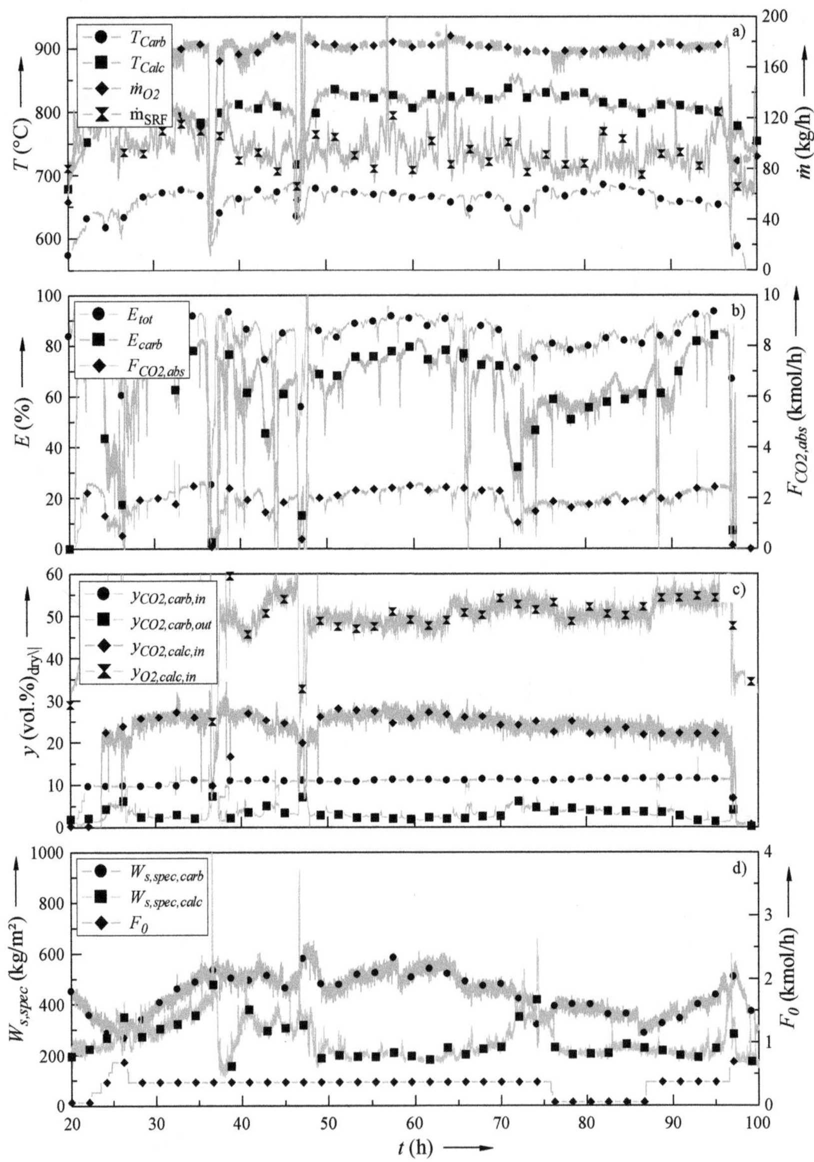

Figure A-2: Data plots (a-c) of relevant process data for the first CaL test campaign (part 2).

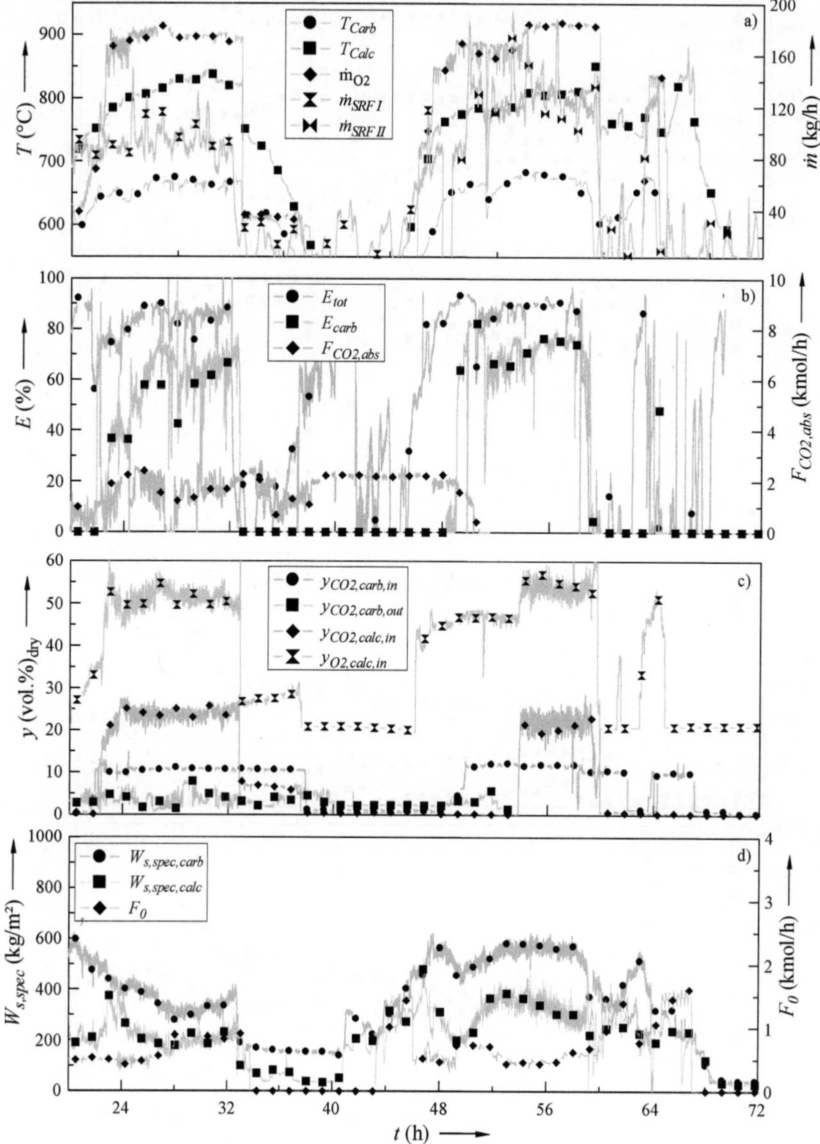

Figure A-3: Data plots (a-c) of relevant process data for the second CaL test campaign (part 1).

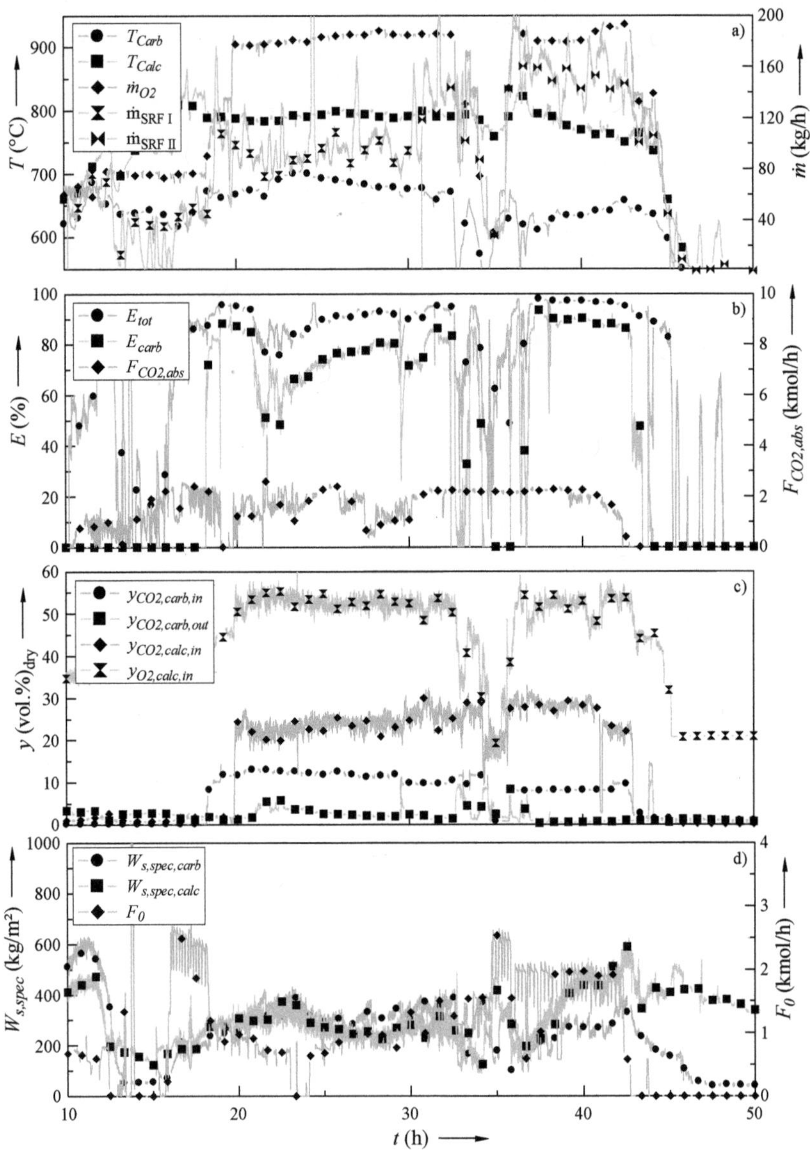

Figure A-4: Data plots (a-c) of relevant process data for the second CaL test campaign (part 2).

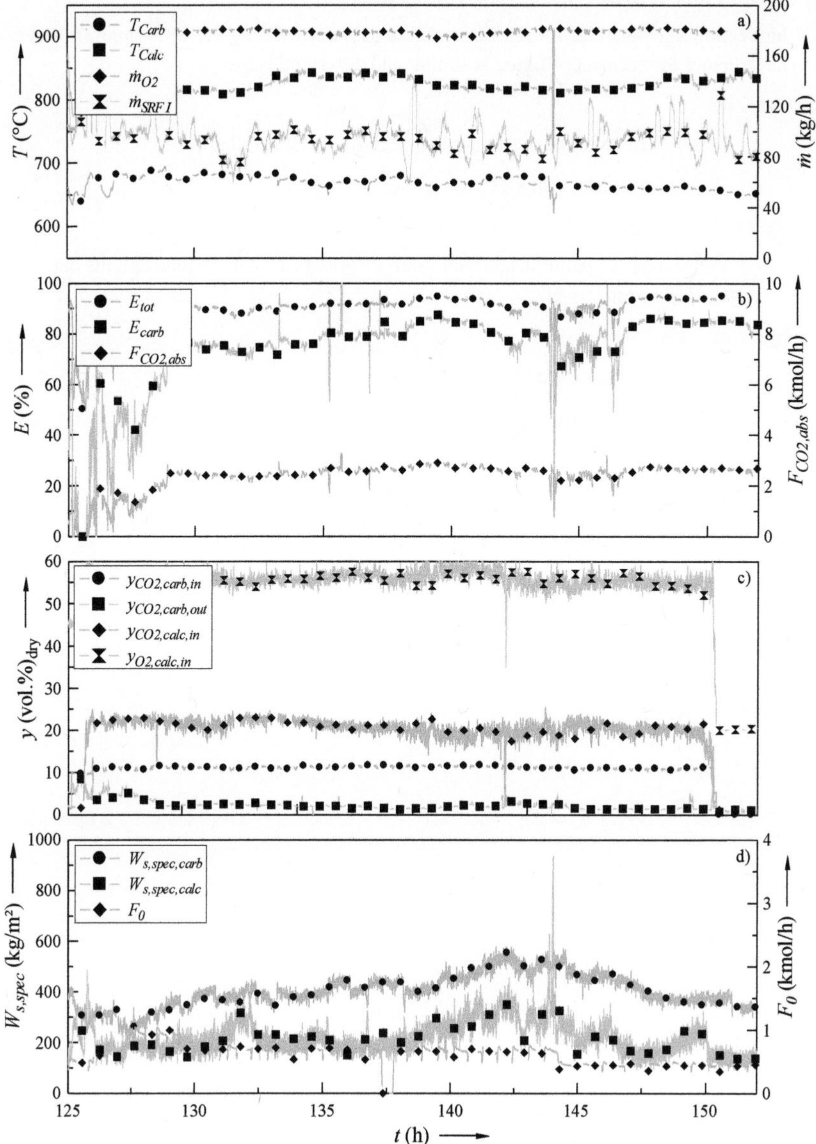

Figure A-5: Data plots of relevant process data for the second CaL test campaign (part 3).

A. 3 Data Plots of SRF Combustion Test Campaign

Within this section, the data plot for relevant process parameter during the SRF combustion tests is given. The structure for each set of figure is similar and described below.

- Figure a)
    - o Left y-axis: Average reactor temperature of the CFB400 (T)
    - o Right y-axis: Mass flow of SRF ($\dot{m}_{SRF}$)
- Figure b)
    - o Left y-axis: Volumetric concentration of O_2 and CO_2 for the inlet and the outlet of the CFB400, respectively.
- Figure c)
    - o Left y-axis: Specific emission of NO (e_{NO})
    - o Right y-axis: Oxygen-to-fuel ratio (λ)
- Figure d)
    - o Left y-axis: Specific emission of HCl (e_{HCl})
    - o Right y-axis: Specific emission of SO_2 (e_{HCl})

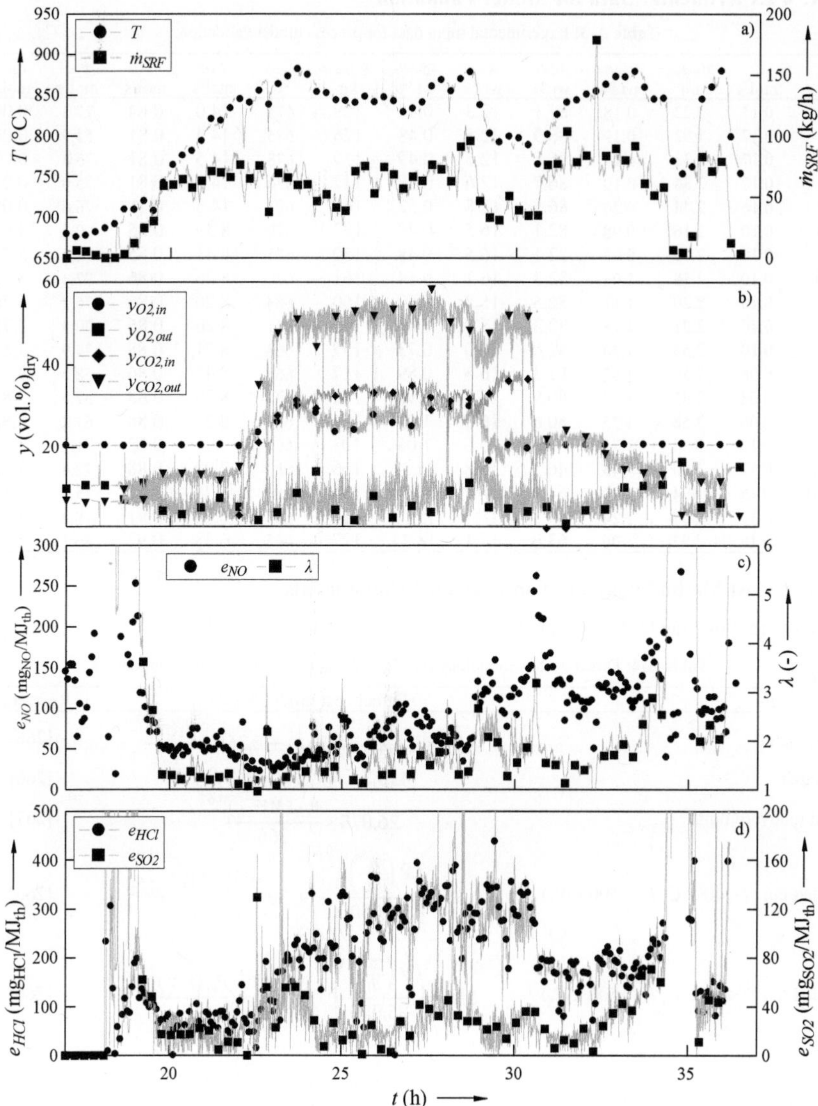

Figure A-6: Data plots of relevant process data for SRF combustion in a stand-alone CFB.

A. 4 Experimental Data for Model Validation

Table A-3: Experimental input data for process model validation.

OP	F_0 mol/s	$\dot{m}_{sorbent}$ t/h	x_{CaSO4} wt.%	x_{CaO} wt.%	x_{Ash} wt.%	x_{CaCO3} wt.%	$W_{S,carb}$ kg	T_{carb} °C	F_{O2} mol/s	F_{CO2} mol/s	F_{N2} mol/s	F_{H2O} mol/s
I	0.17	2.22	0.18	87.1	12.3	0.45	129	672	14.0	0.84	77.3	0.00
II	0.17	1.92	0.19	86.9	12.5	0.48	126	675	14.5	0.81	54.9	0.00
III	0.17	2.33	0.19	86.8	12.5	0.49	139	678	14.5	0.81	73.8	0.00
IV	0.17	2.34	0.19	86.7	12.6	0.50	147	676	14.5	0.81	75.0	0.00
V	0.16	2.44	0.20	86.5	12.8	0.52	166	681	14.5	0.81	76.6	0.00
VI	0.10	2.16	0.98	82.1	16.5	0.48	133	670	8.39	0.85	60.3	3.69
VII	0.10	2.16	0.98	82.1	16.5	0.48	150	671	8.41	0.85	74.1	3.76
VIII	0.10	2.18	1.04	82.3	16.1	0.54	161	670	8.26	0.86	77.4	3.76
IX	0.10	2.20	1.11	82.5	15.8	0.60	160	664	8.20	0.86	76.2	3.76
X	0.10	2.21	1.18	82.3	15.9	0.65	156	661	8.26	0.86	70.3	3.70
XI	0.10	2.53	1.30	82.0	16.0	0.73	137	668	8.21	0.88	74.9	3.83
XII	0.08	2.56	1.52	81.3	16.3	0.89	117	668	8.41	0.86	58.8	3.76
XIII	0.08	2.57	1.64	80.9	16.5	0.98	112	672	8.20	0.88	57.5	3.89
XIV	0.08	2.58	1.75	80.6	16.6	1.05	103	679	8.21	0.88	61.2	3.83
XV	0.10	2.34	1.75	80.6	16.6	1.05	103	660	7.91	0.82	71.1	3.36
XVI	0.12	2.60	1.43	46.2	16.8	35.52	166	682	7.90	0.88	72.4	3.68
XVII	0.19	1.94	1.24	64.1	18.7	15.98	111	680	6.78	0.88	73.2	3.14
XVIII	0.19	2.07	1.24	64.1	18.7	15.98	124	667	6.48	0.91	79.8	3.29
XIX	0.19	2.92	1.06	63.6	17.1	18.24	127	665	6.33	0.92	86.1	3.31

A. 5 Cost Methodology for Non-Standard Components

The costs for non-standard components are derived from the following table.

Table A-4: Direct cost methodology for the evaluation of non-standard components.

Component	Direct cost model, $k_{EUR,2017}$	Reference
Carbonator	$474 * P_{th,EVA\,I}(MW_{th}) + 8{,}360 * \dfrac{d_i(m)^{0.6}}{4.7}$	[266]
Calciner	$415.3 * P_{th,calc}(MW_{th})^{0.65}$	[266]
Heat recovery unit	$26{,}875 * \dfrac{P_{th}(MW_{th})^{0.67}}{266}$	[267]
Sealed fan ($T< 400$ °C, $P_{el} <500$ kW$_e$)	$63 * \left(\dfrac{\dot{V}\left(\frac{m^3}{h}\right)}{25{,}000}\right)^{0.5} + 39 * (\dfrac{P_{el}(kW)}{75})^{0.65}$	[266]
Sealed fan ($T< 400$ °C, $P_{el} >500$ kW$_e$)	$518 * \left(\dfrac{\dot{V}\left(\frac{m^3}{h}\right)}{25{,}000}\right)^{0.5} + 303 * (\dfrac{P_{el}(kW)}{75})^{0.65}$	[266]
ASU	$63{,}645 * (\dfrac{\dot{m}_{O2}\left(\frac{t}{d}\right)}{2{,}717})^{0.6}$	[266]
GPU	$28{,}722 * \dot{m}_{CO2}\left(\dfrac{t}{a}\right)^{0.6}$	[266]

Publications and Presentations

Journal Publications (peer-review)

- **M. Haaf**, R. Anantharaman, S. Roussanaly, J. Ströhle and B. Epple. CO_2 Capture from Waste to Energy Plants: Techno-Economic Assessment of Novel Integration Concepts of Calcium Looping Technology. *Resources, Conservation and Recycling* (accepted for publication).
- **M. Haaf**, J. Hilz, J. Peters, A. Unger, J. Ströhle and B. Epple. Operation of a 1 MW_{th} Calcium Looping Pilot Plant Firing Waste-Derived Fuels in the Calciner. *Powder Technology*, 372:267-274, 2020. DOI: https://doi.org/10.1016/j.powtec.2020.05.074.
- **M. Haaf**, J. Peters, J. Hilz, A. Unger, J. Ströhle and B. Epple. Combustion of Solid Recovered Fuels within the Calcium Looping Process - Experimental Demonstration at 1 MW_{th} Scale. *Experimental Thermal and Fluid Science*, 113(1): 110023, 2020. DOI: 10.1016/j.expthermflusci.2019.110023.
- S. Roussanaly, J.A. Ouassou, R. Anantharaman and **M. Haaf**. Impact of Uncertainties on the Design and Cost of CCS from a waste-to-energy plant. *Frontiers in Energy Research*, 8:17, 2020. DOI: 10.3389/fenrg.2020.00017.
- J. Hilz, **M. Haaf**, M. Helbig, N. Lindqvist, J. Ströhle and B. Epple. Scale-up of the Carbonate Looping Process to a 20 MW_{th} pilot plant based on long-term Pilot Tests. *International Journal of Greenhouse Gas Control*, 88:332-341, 2019. DOI: 10.1016/j.ijggc.2019.04.026.
- **M. Haaf**, P. Ohlemüller, J. Ströhle and B. Epple. Techno-economic assessment of alternative fuels in second-generation carbon capture and storage processes. *Mitigation Adaptation Strategies for Global Change*, 25:149-164 2020. DOI: 10.1007/s11027-019-09850-z.
- A. Rolfe, Y. Huang, **M. Haaf**, A. Pita, S. Rezvani, A. Dave and Neil Hewitt. Technical and Environmental Study of Calcium Carbonate Looping versus Oxy-fuel Options for low CO_2 Emission Cement Plants. *International Journal of Greenhouse Gas Control*, 75:85-97, 2018. DOI: 10.1016/j.ijggc.2018.05.020.
- **M. Haaf**, J. Hilz, M. Helbig, C. Weingärtner, O. Stallmann, J. Ströhle and B. Epple. Assessment of the Operability of a 20 MW_{th} Calcium Looping Demonstration Plant by Advanced Process Modelling. *International Journal of Greenhouse Gas Control*, 75:224-234, 2018. DOI: 10.1016/j.ijggc.2018.05.014.
- A. Rolfe, Y. Huang, **M. Haaf**, S. Rezvani, D. McIlveen-Wright and N. Hewitt. Integration of the Calcium Carbonate Looping Process into an Existing Pulverized coal-fired Power Plant for CO_2 capture: Techno-economic and Environmental Evaluation. *Applied Energy*, 222: 169-179, 2018. DOI: 10.1016/j.apenergy.2018.03.160.
- J. Hilz, M. Helbig, **M. Haaf**, A. Daikeler, J. Ströhle and B. Epple. Investigation of the Fuel Influence on the Carbonate Looping Process in 1 MW_{th} scale. *Fuel Processing Technology*, 169:170-177, 2018. DOI: 10.1016/j.fuproc.2017.09.016.
- F. Alobaid, W. Al-Maliki, T. Lanz, **M. Haaf**, A. Brachthäuser, B. Epple and I. Zorbach. Dynamic Simulation of a Municipal Solid Waste Incinerator. *Energy*, 149:230-249, 2018. DOI: 10.1016/j.energy.2018.01.170
- J. Hilz, M. Helbig, **M. Haaf**, A. Daikeler, J. Ströhle and B. Epple. Long-Term Pilot Testing of the Carbonate Looping Process in 1 MW_{th} Scale. *Fuel*, 210:892-899, 2017. DOI: 10.1016/j.fuel.2017.08.105.

Presentation at International Conferences

- **M. Haaf**, R. Anantharaman, S. Roussanaly, J. Ströhle and B. Epple. CO_2 Capture from Waste to Energy Plants: Techno-Economic Assessment of Novel Integration Concepts of Calcium Looping Technology. *10th Trondheim Conference on CO_2 Capture, Transport and Storage (TCCS-10)*, 17.-19.06.2019, Trondheim, Norway.

- **M. Haaf**, J. Peters, J. Hilz, A. Unger, J. Ströhle and B. Epple. Combustion of Solid Recovered Fuels within the Calcium Looping Process - Experimental Demonstration at 1 MW$_{th}$ Scale. *11th Mediterranean Combustion Symposium (MCS-11)*, 16.-20.06.2019, Tenerife, Spain.
- **M. Haaf**, J. Hilz, J. Peters, A. Unger, J. Ströhle and B. Epple. Oxy-Fuel Combustion of Solid Recovered Fuels in the Fluidized Bed Calciner of a 1 MW$_{th}$ Calcium Looping Unit. *Fluidization XVI conference*, 26.-31.05.2019, Guilin, China.
- J. Hilz, **M. Haaf**, M. Helbig, J. Ströhle and B. Epple, 2018. Scale-up of the Carbonate Looping Process to a 20 MW$_{th}$ Pilot Plant Based on Long Term Pilot Tests. *14th International Conference on Greenhouse Gas Control Technologies (GHGT-14)*, 21.-25.10.2018, Melbourne, Australia.
- **M. Haaf**, J. Hilz, A. Unger, J. Ströhle and B. Epple. Methanol Production via the Utilization of Electricity and CO_2 Provided by a Waste Incineration Plant. *14th Conference on Greenhouse Gas Control Technologies (GHGT-14)*, 21.-25.10.2018, Melbourne, Australian.
- **M. Haaf**, P. Ohlemüller, J. Ströhle and B. Epple, 2018. Assessment of the Potential for Negative CO_2 Emissions by the Utilization of Alternative Fuels in 2nd Generation CCS Processes. *1st International Conference on Negative CO_2 Emissions*, 22.-24.05.2018, Gothenburg, Sweden.
- **M. Haaf**, M. Helbig, J. Hilz, C. Weingärtner, J. Ströhle and Bernd Epple, 2017. Process Design of a 20 MW$_{th}$ Calcium Looping Pilot Plant Based on Long Term Testing in 1 MW$_{th}$ Scale and Advanced Process Modelling. *7th High Temperature Solid Looping Cycle Network Meeting (HTSLCN)*, 04.-05.09.2017, Luleå, Sweden.
- Y. Huang, A. Rolfe, S. Rezvani, **M. Haaf**, A. Dave and N. Hewitt, 2017. Techno-Economic and Environmental Analysis of Calcium Carbonate Looping for CO_2 Capture from a Pulverized Coal-Fired Power Plant. *9th International Conference on Applied Energy (ICAE-9)*, 21.-24.08.2017, Cardiff, United Kingdom.
- **M. Haaf**, M. Helbig, J. Hilz, A. Stroh, J. Ströhle and B. Epple, 2017. Thermodynamic Assessment of the Calcium Looping Process Applied to a Lignite-fired Power Plant. *9th Trondheim Conference on Carbon Capture and Storage (TCCS-9)*, 12.-14.06.2017, Trondheim, Norway.
- M. Helbig, J. Hilz, **M. Haaf**, J. Ströhle and B. Epple, 2017. Evaluation of 1 MW$_{th}$ long-term Pilot Testing of the Carbonate Looping Process. *9th Trondheim Conference on Carbon Capture and Storage (TCCS-9)*, 12.-14.06.2017, Trondheim, Norway.
- **M. Haaf**, F. Borchert, J. Hilz, M. Helbig, A. Stroh, J. Ströhle and B. Epple, 2017. Thermodynamic Assessment of the Calcium Looping Process in Carbon Intense Industries - a Model based Study for a Cement Plant. *9th Trondheim Conference on Carbon Capture and Storage (TCCS-9)*, 12.-14.06.2017, Trondheim, Norway.
- A. Stroh, F. Alobaid, A. Daikeler, **M. Haaf**, M. von Bohnstein, J. Ströhle and B. Epple, 2017. Numerical CFD Simulation of the 1 MW$_{th}$ CFB Carbonator using the Discrete Element Method. *12th International Conference on Fluidized Bed Technology (CFB-12)*, 23.-26.05.2017 Krakau, Poland.
- J. Hilz, M. Helbig, **M. Haaf**, A. Daikeler, J. Ströhle and B. Epple, 2017. Long-term Pilot Testing of the Carbonate Looping Process in 1 MW$_{th}$ Scale. *IEA Clean Coal Centre's 8th International Conference on Clean Coal Technologies (CCT-2017)*, 08.-12.05.2017, Cagliari, Italy.
- M. Helbig, J. Hilz, **M. Haaf**, A. Daikeler, J. Ströhle and B. Epple, 2016. Long-term Carbonate Looping Testing in a 1 MW$_{th}$ Pilot Plant with Hard Coal and Lignite. *13th Conference on Greenhouse Gas Control Technologies (GHGT-13)*, 14.-18.11.2016, Lausanne, Switzerland.
- **M. Haaf**, A. Stroh, J. Hilz, M. Helbig, J. Ströhle and B. Epple, 2016. Process Modelling of the Calcium Looping Process and Validation against 1 MW$_{th}$ Pilot Testing. *13th Conference on Greenhouse Gas Control Technologies (GHGT-13)*, 14.-18.11.2016, Lausanne, Switzerland.

Presentation at National Conferences

- **M. Haaf**, J. Peters, A. Daikeler, V. Barisic, K. Vänskä, A. Unger, J. Ströhle and B. Epple. Experimentelle Untersuchungen zum Einsatz von Ersatzbrennstoffen in wirbelschichtbasierten

Feuerungssystems im 1 MW$_{th}$ Maßstab. *VGB-Fachtagung Thermische Abfallverwertung und Wirbelschichtfeuerungen*, 13.-14.03.2019, Hamburg, Deutschland.

- J. Hilz, **M. Haaf**, A. Daikeler, J. Ströhle and B. Epple, 2019. Carbonate Looping for Industrial Applications. *DECHEMA - Jahrestreffen der ProcessNet-Fachgruppe Energieverfahrenstechnik und des Arbeitsausschusses Thermische Energiespeicherung*, 06. - 04.03.2019, Frankfurt a. Main, Germany.

- J. Hilz, M. Helbig, **M. Haaf**, A. Daikeler, J. Ströhle and B. Epple, 2018. Scale-up of Carbonate Looping Based on Long-term pilot testing in 1 MW$_{th}$ Scale. *DECHEMA - Jahrestreffen der ProcessNet-Fachgruppe Energieverfahrenstechnik*, 07. - 08.03.2018, Frankfurt a. Main, Germany.

- J. Hilz, M. Helbig, **M. Haaf**, A. Daikeler, J. Ströhle and B. Epple, 2017. CO_2 Capture by means of limestone - Calcium Carbonate Looping. *DECHEMA - Jahrestreffen der ProcessNet-Fachgruppen Abfallbehandlung und Wertstoffrückgewinnung, Energieverfahrenstechnik, Gasreinigung, Hochtemperaturtechnik, Rohstoffe und Kreislaufwirtschaft*, 21. - 23.03.2017, Frankfurt a. Main, Germany.

Other Publications

- **M. Haaf,** A. Müller, A. Unger, J. Ströhle and B. Epple. Combustion of Solid Recovered Fuels in a Semi-Industrial Circulating Fluidized Bed Pilot Plant - Implications of Bed Material and Combustion Atmosphere on Gaseous Emissions. *VGB PowerTech*, 3(2020): 51-56.

- **M. Haaf**, A. Unger, J. Ströhle and B. Epple. Erzeugung von Methanol aus CO_2 und Elektrizität an einer Müllverbrennungsanlage. *BWK Das Energie Fachmagazin*, 71(9):60-63, 2019.

- J. Hilz, **M. Haaf**, M. Helbig, J. Ströhle and B. Epple. Calcium Carbonate Looping: CO_2 capture by using limestone in the cement industry. *Cement International*, 15:52-63, 2017.